A. Minguzzi – S. Succi – F. Toschi
M.P. Tosi – P. Vignolo

Numerical methods for atomic quantum gases

APPUNTI

SCUOLA NORMALE SUPERIORE
2004

ISBN: 88-7642-130-0

Numerical methods for atomic quantum gases

A. Minguzzi [a], S. Succi [b], F. Toschi [b], M.P. Tosi [a] and
P. Vignolo [a]

[a] *NEST-INFM and Scuola Normale Superiore, Piazza dei Cavalieri 7, I-56126
Pisa, Italy*

[b] *Istituto per le Applicazioni del Calcolo, CNR, Viale del Policlinico 137, I-00161
Roma, Italy*

Abstract

The achievement of Bose-Einstein condensation in ultra-cold vapours of alkali atoms
has given enormous impulse to the study of dilute atomic gases in condensed quantum states inside magnetic traps and optical lattices. High purity and easy optical
access make them ideal candidates to investigate fundamental issues on interacting
quantum systems. This review presents some theoretical issues which have been
addressed in this area and the numerical techniques which have been developed and
used to describe them, from mean-field models to classical and quantum simulations for equilibrium and dynamical properties. After an introductory overview on
dilute quantum gases, both in the homogeneus state and under harmonic or periodic confinement, the article is organized in three main chapters. The first concerns
Bose-condensed gases at zero temperature, with main regard to the properties of the
ground state in different confinements and to collective excitations and transport in
the condensate. Bose-Einstein-condensed gases at finite temperature are addressed
in the next chapter, the main emphasis being on equilibrium properties and phase
transitions and on dynamical and transport properties associated with the presence
of the thermal cloud. Finally, the last chapter is focused on theoretical and computational issues that have emerged from the efforts to drive gases of fermionic atoms
and boson-fermion mixtures deep into the quantum degeneracy regime, with the
aim of realizing novel superfluids from fermion pairing. The attention given in this
article to methods beyond standard mean-field approaches should make it a useful
reference point for future advances in these areas.

CONTENTS

1 Overview on dilute quantum gases

1.1 Introduction

Bose-Einstein condensation can occur when a gas of bosonic atoms is cooled down to the point where the de Broglie wavelength $l_B = \hbar/m v_T$ becomes comparable with the mean interparticle separation $d = n^{-1/3}$. Here v_T is the atomic thermal speed, m is the atomic mass, and n is the atomic number density. Under these conditions the atomic wave packets overlap and quantum interference between identical particles becomes crucial in determining the statistical behaviour of the gas. A phase transition leads to the formation of a Bose-Einstein condensate (BEC), namely to a coherent cloud of atomic matter in which a macroscopic number of atoms occupy the same quantum ground state, thereby forming a sort of "giant matter wave".

The transition temperature and the peak atomic density in an ideal gas are related as

$$n l_B^3 \sim 2.612 . \tag{1}$$

At a typical density $n = 10^{12} - 10^{14}$ particles per cc, the BEC starts to form in alkali gases at temperatures around a hundred nK, which is seven orders of magnitude lower than for the superfluid state in liquid ^{4}He. At such ultra-low temperatures classical interactions would localize the atoms and prevent quantum overlaps, on a time scale which is dictated by binary and higher collisions leading to the formation of clusters and superatomic aggregates. In the experiments yielding a BEC, however, the gas is cooled along an out-of-equilibrium path which quenches the kinetic energy of the atoms while they are still in the gaseous state. Three-body and higher collisions are the main processes which limit the lifetime of the condensate to a few seconds, but this is long enough to perform experiments in the metastable state under quasi-equilibrium conditions.

In current experiments the gas is subjected to an external potential due to the magnetic or optical fields that are used to cool and hold the atoms. The final output of the cooling process is a mesoscopic phase-coherent droplet of micrometer size. The inhomogeneity due to the external confinement deeply affects the physical properties of the gas, such as its spatial distribution (with a spectacular condensate peak appearing at the trap centre), its thermodynamic laws, its excitation spectrum, and its behaviour at the phase transition boundary. The introduction of a new length and energy scale allows for new equilibrium states that are not available in the homogeneous macroscopic limit and modifies the role of fluctuations.

A very attractive feature of BEC's is that they are amenable to simple and yet quite basic theoretical approaches. The theory was developed under the assumption that binary collisions occur much more frequently than higher-order collisions. This is the well known dilute-gas approximation, which lies at the heart of classical and quantum kinetic theory. The diluteness condition is fulfilled whenever the mean interparticle distance is much larger than the range of the interatomic forces. For quantum gases the latter can be identified with the s-wave scattering length a, which is typically in the range of a few nanometers, so that the diluteness parameter is very small indeed, $\sqrt{na^3} \sim 10^{-3}$. Under such conditions the interatomic forces can be modelled by contact interactions of the form $V(\mathbf{r}, \mathbf{r}') = g\delta(\mathbf{r} - \mathbf{r}')$, with

$$g = \frac{4\pi\hbar^2 a}{m}. \tag{2}$$

This interaction contains the exact low-energy scattering amplitude in the Born approximation. Positive a indicates repulsion, and in the experiments both positive or negative a are met.

In the dilute limit each atom feels the effects of all surrounding atoms to first (mean-field) approximation *via* an effective potential energy which is proportional to the local density of the condensate, and the depletion of the condensate due to quantum fluctuations is very small as it scales with the dilution parameter $\sqrt{na^3}$. In these conditions the confined gas at zero temperature is fully characterized by the condensate wavefunction, which keeps trace of the gas density and of the phase of the condensate. The condensate wavefunction is determined by a nonlinear Schrödinger equation known as the Gross-Pitaevskii equation (GPE). The GPE plays a pivotal role in the study of dilute BEC's at low temperature, since it describes the ground state as well as linear and non-linear transport phenomena. The GPE however is not able to treat strongly correlated systems, as can be the case of bosons in reduced dimensionality or with long-range interactions or near a Feshbach resonance. In these cases one may resort to numerical simulations.

At finite temperature the GPE needs to be coupled with dynamical equations for the thermal excitations. The standard mean-field approach to the thermal cloud is based on the Hartree-Fock-Bogoliubov (HFB) theory, in which the condensate and noncondensate atoms interact *via* a temperature-dependent mean field. Another approach, which is most suited to higher temperature, is to represent the thermal excitations *via* their one-body kinetic distribution function $f(\mathbf{r}, \mathbf{p}, t)$ (the Wigner distribution, related by a Fourier transform to the one-body density matrix). Collisions between atoms in the condensate and thermal atoms can also be included consistently in the theory and in simulations of kinetic equations. As an alternative Path-Integral Monte Carlo simulations have been developed specifically for dilute gases.

Dilute Fermi gases are also being trapped and cooled in several laboratories.

In the fermionic case degeneracy is reached by populating a large set of single-particle levels inside the Fermi sphere, but for attractive effective interactions a superfluid phase can arise through pairing of fermions. Understanding the role of fermionic statistics and the properties of the superfluid phase has been the subject of extensive theoretical investigations. Bose-Fermi mixtures have also been studied, in connection with the experimental strategy of sympathetic cooling of fermions by s-wave collisions with bosons. A rich phase diagram and possible new phases have been predicted for these mixtures. A semiclassical approach to treat the dynamics of interacting fermions within a Vlasov-Landau picture has been set up. Fully quantum simulations are notoriously much more difficult for fermionic systems than for their bosonic counterparts, since the sign problem hampers the interpretation of the density matrix as a probability distribution.

As experiments keep probing BEC and ultracold Fermi gases under an increasing variety of situations, substantial theoretical efforts are being spent to understand and characterize the experiments as well as in investigating fundamental questions. Much of this theoretical effort is conducted by analytical means, with the whole array of tools of quantum statistical mechanics. As is always the case with complex physical phenomena, analytical investigation faces severe limitations whenever genuinely non-perturbative and strongly nonlinear effects need to be quantitatively addressed. Under these circumstances help from numerical investigations becomes mandatory. The central aim of the present work is to address the main aspects related to the numerical study of Bose-Einstein condensates and of atomic Fermi gases.

Due to the enormous breadth of the subject, no review can possibly hope to give a full coverage of the ground. A series of reviews and books has already covered many of the experimental and theoretical issues which have emerged in the first few years after the achievement of Bose-Einstein condensation (see Parkin and Walls [1], Dalfovo et al. [2], Leggett [3], Pethick and Smith [4], Pitaevskii and Stringari [5], and the schools held at Varenna [6], Erice [7], Cargèse [8,9] and Les Houches [10]). In this review we focus on the numerical techniques which have been developed for atomic gases both for solving the mean-field problem and for studying the strongly correlated regimes. We shall try to illustrate the major directions developed so far and to offer some perspective remarks on future directions.

1.2 Homogeneous state

Let us start by considering a model system made of N spinless Bose particles inside a macroscopic box and interacting with short-range forces. Such a model was first introduced in the attempt to describe the physics of ultracold ^{4}He.

The idea due to Bogoliubov is to perform a perturbative expansion for a dilute real gas, taking into account that a macroscopic number of bosons lie in the condensate in the zero-momentum state. The Bogoliubov expansion and the more refined treatments which have followed it were not able to give a quantitative description of liquid ^{4}He, for which $\sqrt{na^3} \simeq 1$, but are an excellent starting point for atomic gases.

1.2.1 Scattering length

We describe the gas by the Hamiltonian

$$\mathcal{H} = \sum_{\mathbf{k}} \varepsilon_{\mathbf{k}} a_{\mathbf{k}}^{\dagger} a_{\mathbf{k}} + \frac{1}{2} \sum_{\mathbf{q}} \sum_{\mathbf{k},\mathbf{k'}} v_{\mathbf{q}} a_{\mathbf{k}-\mathbf{q}}^{\dagger} a_{\mathbf{k'}+\mathbf{q}}^{\dagger} a_{\mathbf{k'}} a_{\mathbf{k}} \tag{3}$$

where $a_{\mathbf{k}}$ and $a_{\mathbf{k}}^{\dagger}$ are the annihilation and creation operators of a particle with momentum $\hbar\mathbf{k}$, the free particle excitation energy is $\varepsilon_{\mathbf{k}} = \hbar^2 k^2/2m - \mu$ relative to the chemical potential μ, and the Fourier transform of the interparticle potential is $v_{\mathbf{q}}$. In all calculations on dilute gases it is customary to adopt for $v_{\mathbf{q}}$ a Fermi pseudopotential $v_{\mathbf{q}} = 4\pi\hbar^2 a/m$, corresponding to a contact interaction parametrized by an s-wave scattering length a. This choice is dictated by several considerations: (*i*) the true interatomic potential has many bound states, reflecting the fact that the true ground state is a solid and not an atomic vapour, and these bound states would lead to instabilities in the actual calculations; (*ii*) the typical interparticle collisions at ultra-low temperatures are only in the s-wave channel, since higher partial waves are suppressed by centrifugal effects [11]; and (*iii*) the above choice for $v_{\mathbf{q}}$ implicitly takes into account the repeated scattering processes between each pair of particles (see *e.g.* the book of Abrikosov *et al.* [12]).

The homogeneous gas is stable for repulsive interactions ($a > 0$). Of course, the contact-potential model fails at large $\mathbf{q}$, as the detailed structure of the colliding atoms becomes relevant at very small distances, of the order of the atomic effective radius. This failure is reflected in ultraviolet divergencies that are met in calculations beyond mean field if one does not take into account all relevant terms in the s-wave scattering amplitude [13].

The above discussion applies also to a confined gas only as long as the typical size of the gaseous cloud is much larger than the scattering length. In this case one can safely employ the pseudopotential with the value of the scattering length as derived in the absence of the trap. A new two-particle scattering problem needs to be solved when the scattering length becomes comparable with the trap size, and the presence of the confinement may drastically change the physical picture. Thus in quasi-2D gases the magnitude and sign of the scattering amplitude can depend on external system variables [14] and in quasi-1D gases the scattering amplitude vanishes in the strong-coupling limit [15].

1.2.2 The Bogoliubov approximation

If a macroscopic number N_0 of particles are in the condensate state $|\mathbf{k} = 0\rangle$, then one can neglect the commutator $[a_0^\dagger, a_0]$ and set $a_0 \simeq \sqrt{N_0}$ and $a_0^\dagger \simeq \sqrt{N_0}$. This is the essence of the so-called Bogoliubov position [16], which introduces a new set of field operators $b_\mathbf{k}$ defined through $a_\mathbf{k} = \sqrt{N_0}\delta(\mathbf{k}) + b_\mathbf{k}$ and retains in the Hamiltonian (3) only terms up to quadratic order in $b_\mathbf{k}$ and $b_\mathbf{k}^\dagger$. The resulting Hamiltonian $\mathcal{H}_B$ can be cast in a diagonal form by a canonical transformation, leading to

$$\mathcal{H}_B = E_g + \sum_\mathbf{k} \hbar\omega_\mathbf{k} \alpha_\mathbf{k}^\dagger \alpha_\mathbf{k} \tag{4}$$

for free quasiparticles described by the operators $\alpha_\mathbf{k}$ and $\alpha_\mathbf{k}^\dagger$, which are linear combinations of $b_\mathbf{k}$ and $b_{-\mathbf{k}}^\dagger$. Here E_g is the ground-state energy and the excitation energy of a quasiparticle is

$$\hbar\omega_\mathbf{k} = \sqrt{(2gn_0 + \xi_\mathbf{k})\,\xi_\mathbf{k}} \tag{5}$$

where $\xi_\mathbf{k} = \hbar^2 k^2/2m$.

At long wavelengths the single-particle Bogoliubov spectrum has linear dispersion,

$$\omega_\mathbf{k} \to ck \tag{6}$$

with $c = \sqrt{gn_0/m}$. This is similar to the dispersion relation for the collective (sound-wave) density fluctuations. The coincidence of the single-particle and collective excitation spectra at long wavelengths is indeed always valid at zero temperature in the presence of a Bose-Einstein condensate, as was demonstrated by Gavoret and Nozières [17] (see Sec. 1.2.3). This fact can be exploited in practice to probe the single-particle properties by applying a density modulation to the condensate (see Sec. 1.3.2).

Within the Bogoliubov framework the condensate fraction at zero temperature and the ground-state energy can be estimated as

$$N_0/N = 1 - (8/3\sqrt{\pi})\sqrt{na^3} \tag{7}$$

and

$$E_g/N = \frac{1}{2}gn\left[1 + (128/15\sqrt{\pi})\sqrt{na^3}\right] \tag{8}$$

(for a derivation see e.g. [12]). The effect of the repulsive interactions is to induce a quantum depletion of the condensate and to increase the value of the ground-state energy per particle over the mean field result $gn/2$.

Finally, it may be remarked that the Bogoliubov theory has also been formulated in a number-conserving fashion without the need of neglecting the commutator $[a_0^\dagger, a_0]$ [18].

1.2.3 Rigorous results for the excitation spectrum

While the general problem of a 3D interacting Bose gas cannot be solved without the aid of numerical methods, some rigorous results have been demonstrated for the excitation spectrum in the long wavelength limit, and we report them as they can serve for tests of approximate and numerical treatments.

The Hugenholtz-Pines relation [19] ensures that the single-particle excitation spectrum is gapless in the presence of a condensate: the dispersion relation for the excitations starts out from the chemical potential. Formally, gaplessness is embodied in a constraint between the chemical potential and the static limit of the self-energies. As can be verified in the Bogoliubov approximation, the low-energy excitations are due to fluctuations of the phase of the condensate and the absence of a gap in the excitation spectrum at long wavelength is a general consequence of the Goldstone theorem. In a Bose-condensed system spontaneous breaking of gauge symmetry has occurred, since the phase of the condensate could *a priori* take any value, and phase fluctuations of long wavelength do not require an energy cost.

The Gavoret-Nozières theorem [17] ensures that in the presence of a condensate the spectra of single-particle and collective excitations share the same poles at zero temperature. Such hybridization between the two spectra can be understood qualitatively in a two-fluid picture viewing the Bose fluid as a "mixture" of a superfluid component and a normal component (see Hohenberg and Martin [20] for a rigorous proof in the hydrodynamic region). Due to the presence of the condensate the system admits an additional hydrodynamic mode which is associated with superfluid motions. Since all particles in the condensate share the same quantum state, this mode oscillates with the frequency typical of single-particle motions, but at the same time contributes to the total current and to the density fluctuations of the fluid.

The superfluid velocity $\mathbf{v}_s$ is defined through the gradient of the phase of the condensate,

$$\mathbf{v}_s = \hbar \nabla \phi / m. \qquad (9)$$

The crucial point is that the superfluid velocity field is irrotational, so that the shear viscosity resides entirely in the normal component of the fluid.

While for a weakly interacting Bose gas at rest the phenomenon of Bose-Einstein condensation is related to the macroscopic population of the single-particle state at zero momentum, a more general definition has been given by Penrose and Onsager [21], who associated the BEC with the emergence of long-range "off-diagonal" order in the one-body density matrix $\rho(\mathbf{x}, \mathbf{x}') = \langle \hat{\Psi}^\dagger(\mathbf{x}) \hat{\Psi}(\mathbf{x}') \rangle$. Here $\hat{\Psi}(\mathbf{x})$ and $\hat{\Psi}^\dagger(\mathbf{x})$ are the annihilation and creation field operators for a particle at position $\mathbf{x}$ and $\langle \ldots \rangle$ denotes the average over a suitable ensemble taking into account the presence of the condensate (see *e.g.* Hohenberg and Martin [20]). In this framework the condensate density n_0 is given by

$$n_0 = \lim_{|\mathbf{x}-\mathbf{x}'|\to\infty} \rho(\mathbf{x} - \mathbf{x}'). \tag{10}$$

Since the Fourier transform of $\rho(\mathbf{x} - \mathbf{x}')$ yields the one-body momentum distribution $n(\mathbf{k})$, Eq. (10) can be restated as

$$n(\mathbf{k}) = n_0 \delta(\mathbf{k}) + \tilde{n}(\mathbf{k}). \tag{11}$$

That is, the condensate contributes to $n(\mathbf{k})$ with a peak at zero momentum, while the remainder $\tilde{n}(\mathbf{k})$ is due to the particles out of the condensate.

The total particle density n is given by

$$n = \lim_{|\mathbf{x}-\mathbf{x}'|\to 0} \rho(\mathbf{x} - \mathbf{x}') \tag{12}$$

and in a superfluid the density n_s of the superfluid component becomes equal to n as $T \to 0$. The two macroscopic densities n_0 and n_s thus explore the density matrix in entirely different domains, so that Bose-Einstein condensation and superfluidity are two distinct concepts. The idea of superfluidity was introduced to describe the peculiar transport behaviour of liquid ^{4}He in the II-phase below the λ line, including non-viscous flow through thin capillaries [22], propagation of heat waves [23], and existence of persistent currents [24]. In the two-fluid model of Tisza [25] and Landau (see Khalatnikov's book [26]) the liquid is viewed as if it were a "mixture" of two fluids, a normal fluid which possesses Newtonian viscosity and a superfluid which is capable of frictionless flow around obstacles. Each component is characterized by its density and its velocity field. The superfluid density has been measured as a function of temperature in He-II by recording the period of torsional oscillations, and hence the moment of inertia, of a pile of thin metal disks that are closely spaced so as to ensure that the normal fluid in the interstices would be dragged along in their rotation while the superfluid remains stationary [27] (see also the book by Atkins [28]). It was found that the superfluid fraction increases from zero at the λ point to essentially unity at 1 K (see Fig. 1). In contrast, the condensate fraction is depleted by the effect of the interactions and is only about 7% at the lowest temperatures [29–34] (see Fig. 2). A mesoscopic analogue of the

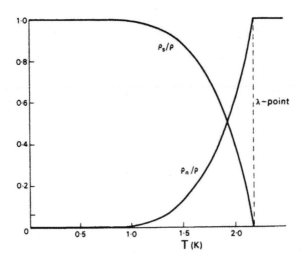

Fig. 1. Superfluid fraction in liquid ^{4}He as obtained in the Andronikashvili experiment.

Andronikashvili experiment has been realized by studying the rotations of a molecule trapped inside a Helium droplet [35–37].

A superfluid set into rotation at high enough angular velocity becomes threaded by vortex lines. The irrotationality of the superfluid component implies that its density must vanish at the vortex core and that the vortex lines are quantized according to the Onsager-Feynman condition

$$\oint \mathbf{v}_s \cdot d\mathbf{l} = nh/m, \tag{13}$$

where n is an integer number and $d\mathbf{l}$ is an element of a closed circuit embracing the vortex line. Equation (13) is an immediate consequence of Eq. (9) in the presence of a condensate, since the phase of the order parameter must be single-valued *modulo* 2π.

Various microscopic definitions can be given for the superfluid density n_s. The first, due to Josephson [38] and Bogoliubov (see *e.g.* Popov's book [39]), relates it to the static limit of the superfluid-velocity autocorrelation function,

$$\lim_{\mathbf{k}\to 0}\lim_{\omega\to 0} \operatorname{Re} \chi_{v_s v_s}(\mathbf{k}, \omega) = \frac{1}{mn_s}. \tag{14}$$

By recalling that the superfluid current is $\mathbf{J}_s = n_s\mathbf{v}_s$ we immediately see that Eq. (14) is the f-sum rule for the superfluid component. From the relation

10

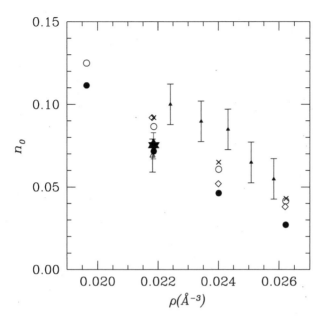

Fig. 2. Condensate fraction in liquid ^{4}He as a function of density, from neutron scattering experiments (at 0.75 K [30] (solid triangles) and extrapolated to $T = 0$ [29] (star)) and from various numerical simulations [31–34] (other symbols). Adapted from Moroni *et al.* [31].

between the superfluid velocity and the gradient of the phase of the order parameter, Eq. (14) can also be expressed in terms of the single-particle Green's functions for the Bose fluid [39].

A second definition of superfluid density is in terms of the transverse-current autocorrelations [40],

$$\lim_{k \to 0} \lim_{\omega \to 0} \operatorname{Re} \chi_{JJ}^T(\mathbf{k}, \omega) = m(n - n_s). \tag{15}$$

That is, only the normal component responds to a transverse external probe since the superfluid component is irrotational.

In Path-Integral Monte Carlo simulations the growth of the superfluid component with decreasing temperature is associated with the growth of many-boson exchange processes in the one-body density matrix (see Sec. 3.1.2). Since there is no clear relation between superfluidity and Bose-Einstein condensation at such a level, we may conclude with Leggett [41] that they both are consequences of deeper topological properties of the many-body wavefunction. It is indeed possible to have superfluid behaviour even in the absence of a true condensate, as is the case in low-dimensional Bose fluids.

1.2.5 Low-dimensional Bose gases

Phase fluctuations are enhanced in low-dimensional Bose fluids. Already in an ideal Bose gas there is no true macroscopic condensate at any finite temperature in 2D and no condensate ever in 1D.

More generally, the absence of off-diagonal long-range order at finite temperature T in 2D and 1D Bose superfluids was demonstrated by Hohenberg [42] by reasoning *ad absurdum*. The presence of the condensate would lead to the inequality

$$\tilde{n}(\mathbf{k}) \geq -\frac{1}{2} + \frac{mk_BT}{\hbar^2k^2}\frac{n_0}{n} \tag{16}$$

for the momentum distribution at low momenta, but for dimensionality $d \leq 2$ this implies a divergence of the total particle density, which is given by $n = \int d^dk\, n(\mathbf{k})$. One is led to the conclusion that the condensate fraction must be zero at $T \neq 0$.

In 2D Bose fluids at $T = 0$ the picture depends on the type of interactions. A gas with short-range interactions or $1/r$ repulsions shows condensation, but a charged fluid interacting with the $\ln r$ potential does not condense even at $T = 0$ [43]. The appropriate inequality is

$$\tilde{n}(\mathbf{k}) \geq -\frac{1}{2} + \frac{1}{S(k)}\frac{n_0}{n} \tag{17}$$

for $\tilde{n}(\mathbf{k})$ in terms of the structure factor $S(k)$. This has been derived by Pitaevskii and Stringari [44] through a generalization of the uncertainty principle and was used by them to prove that off-diagonal long-range order is absent in several 1D systems with continuous-group symmetries. From the inequality (17) Magro and Ceperley [43] have shown that plasma density fluctuations destroy the condensate in the 2D fluid with $\ln(r)$ interactions: in this system the structure factor at low momenta is $S(k) \propto k^2/\Omega$ with $\Omega = (2\pi ne^2/m)^{1/2}$ being the plasma frequency, so that again n would diverge if $n_0 \neq 0$. The same argument does not hold for the Bose plasma with $1/r$ interactions, where $\Omega \propto k^{1/2}$ and $S(k) \propto k^{3/2}$, nor for a neutral Bose gas where $S(k) = \hbar k^2/(2m\omega_\mathbf{k}) \propto k$.

The 2D Bose fluid with short-range interactions, on the other hand, undergoes at finite temperature a transition of the Kosterlitz-Thouless type into a superfluid state [45,39]. This state is characterized by local phase coherence and is called a "quasicondensate". At low temperature the one-body density matrix decays to zero with a power-law [39], in contrast to a classical gas where the decay is exponential and to a true condensate where the decay is to a finite asymptotic value n_0 (see Eq. (10)). In practice, this means that the fluid behaves like a genuine condensate if the size of the sample is smaller than the phase coherence length.

Ultracold atomic gases in a metastable state are realized in the laboratory inside traps made from magnetic or optical fields. Such inhomogeneous confinement is usually well approximated by an external potential obeying a harmonic or a sinusoidal law, and is a key ingredient in the characterization of the gas. Current traps are extremely versatile: they can be tuned in space to vary the geometry of the confining potential and periodically varied in time, or suddenly turned off to allow free expansion of the gas. The interplay between external potential and atom-atom interactions gives rise to a variety of new physical effects which are amenable to observation.

1.3.1 The quest for BEC

The realization of ultracold atomic gases originated as an application of very precise and well controlled laser beams, thus providing an alternative to the usual buffer-gas technique of cooling by contact with a cold reservoir [46].

The first proposal for "laser-cooling" of free atoms was based on the Doppler effect [47]. An atom moving in a weak standing wave, which is slightly detuned to the red away from an atomic transition, absorbs a photon from the counterpropagating laser wave and re-emits it by spontaneous emission in a random direction. The atom slows down and the final temperature T_D, with the meaning of an average kinetic energy, is determined by the natural width of the excited state ($T_D \simeq 140\mu K$ for ^{87}Rb). Other laser-cooling mechanisms that lead to temperatures well below the Doppler limit have later been proposed and realized. The so-called Sisyphus cooling [48] uses a laser polarization gradient which removes the degeneracy of the atomic ground-state sublevels as a function of the spatial position: the atom is pumped from a sublevel to another as it moves in space and loses kinetic energy. The temperature limit T_R for such a mechanism is determined by the recoil energy in the emission of a photon ($T_R \simeq 0.3\mu K$ for ^{87}Rb). This limit has been overcome for a gas of metastable ^{4}He atoms by using a laser-cooling scheme based on a velocity-selective optical pumping of atoms into a non-absorbing coherent superposition of states [49]. However, such a scheme is impracticable for atoms with a richer internal structure [50] and cannot be applied to high-density spatial distributions since collisions drive the atoms out of the non-absorbing state [51].

Laser cooling of the gaseous cloud is followed by evaporative cooling inside a magnetic trap [52]. This consists of progressively eliminating the "hot" atoms from the trap while allowing thermalization of the remaining atoms *via* elastic collisions. The cooling process competes with heating due to losses, which are

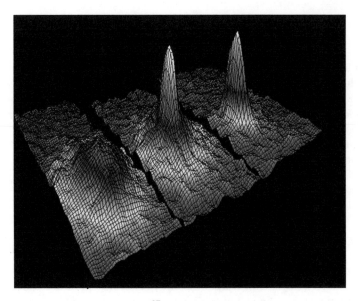

Fig. 3. The first BEC in a gas of ^{87}Rb atoms, as it appears in two-dimensional time-of-flight images as temperature is decreased below T_c. From Anderson *et al.* [53].

mainly in the three-body channel close to the condensation region, at relatively high densities. The condition for Bose-Einstein condensation, as expressed in terms of the phase-space density $n l_{dB}^3 \simeq 2.6$ is extremely severe: for alkali atoms it requires reaching temperatures $T \sim 0.1\mu K$ and densities $n \sim 10^{13}$ atoms/cc.

After the evaporative cooling cycle the condensate appears as a high-density peak at the centre of the density distribution of the atomic cloud, which can be imaged by a resonant absorption technique after release from the trap and ballistic expansion (see Fig. 3). Due to the inhomogeneity of the confinement it is thus possible to observe "condensation" also in real space: roughly speaking, the condensate populates macroscopically the lowest level of the harmonic trap, whose wavefunction is localized around the trap centre. A detailed account of the route to BEC can be found in the 2001 Nobel Lectures [54,55].

At the time of writing Bose-Einstein condensation of atomic gases has been achieved in more than forty laboratories around the world. In addition to the originally condensed species (^{87}Rb [53], ^{23}Na [56], and ^{7}Li [57]) condensation has been reported in gases of spin-polarized H [58], of ^{85}Rb [59], of metastable ^{4}He [60,61], of ^{39}K [62], of ^{133}Cs [63], of ^{172}Yb and ^{176}Yb [64].

14

1.3.2 Characterization of the condensate

The first experiments and theoretical studies have been devoted to confirming that the peak in the time-of-flight images is due to a BEC and to explore its main thermodynamic and dynamical properties.

(i) Density profiles. The 2D absorption image of the Bose-condensed cloud after expansion shows an anisotropic density distribution, which is determined by the anisotropy of the trap. If one assumes that all atoms in the condensate can be described by a single macroscopic wavefunction $\Phi(\mathbf{r})$, which is the ground state in the potential well, then also the momentum distribution $n(\mathbf{k}) = |\int d^3r \exp(i\mathbf{k} \cdot \mathbf{r})\Phi(\mathbf{r})|^2$ of the condensate is anisotropic. The density profile of the cloud after ballistic expansion thus reflects its momentum distribution before expansion.

A consequence of the confinement is that the role of the interactions is magnified. Even in a very dilute gas (typically $\sqrt{n(0)a^3} \sim 10^{-3}$, where $n(0)$ is the top density at the centre of the trap) the shape and size of the condensate are fixed by the interactions. Assuming repulsive contact interactions and harmonic confinement with trap frequency ω, the mean size R of the condensate can be estimated by balancing the confinement energy $\sim m\omega^2 R^2/2$ and the interaction energy $E_{int} \sim gN/R^3$. This estimate yields a radius $R \sim (Na/a_{ho})^{1/5}a_{ho}$, which can be much larger than the bare harmonic oscillator length $a_{ho} = \sqrt{\hbar/m\omega}$. In the early condensates the scaling parameter $\eta = (Na/a_{ho})^{1/5}$ was typically of order 10. The kinetic energy near the centre of the trap is negligible, since it scales as $E_{kin} \sim \hbar^2/(mR^2)$ and thus $E_{kin}/E_{int} \sim \eta^{-4}$. In practice the kinetic energy density becomes appreciable only in the outer regions of the Bose-condensed cloud, where it determines a smoothly vanishing profile.

(ii) Thermodynamic properties. The number of particles in the condensate can be controlled in the cooling process, so that it is possible to investigate the thermodynamic properties of the Bose gas from the thermal (almost classical) regime down to a practically pure condensate [65]. In the experiments the temperature is extracted by fitting the "thermal" tails of the density distribution and the condensate fraction is obtained from the relative numbers of atoms that contribute to the central peak and to the tails. The mean internal energy, excluding the contribution from the confinement, is obtained from the rate of expansion of the cloud in the ballistic regime.

While in a strict sense a finite system cannot undergo sharp phase transitions, nevertheless the Bose-Einstein condensation transition is quite sharp in a mesoscopic cloud with a typical number of atoms of 10^4 to 10^6. The thermodynamic properties of the gas under confinement are qualitatively different from the homogeneous case, since the density of states is strongly dependent

15

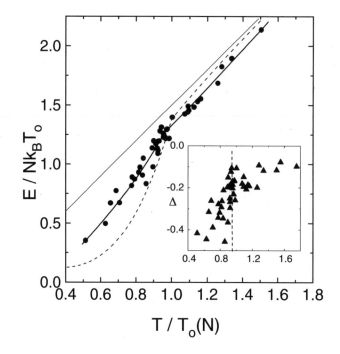

Fig. 4. Measured internal energy of a gas of ^{87}Rb atoms as a function of temperature (symbols), compared with results for an ideal Bose gas (dashed line) and for a classical gas (straight line). The solid lines through the data are polynomial fits of the regions below and above the empirical transition temperature $T_c = 0.94T_0$. The inset shows the deviations from the classical-gas energy in the region near the critical temperature. From Cornell *et al.* [67].

on the confinement. The condensate fraction for an ideal gas in an isotropic harmonic trap is $N_0/N = 1 - (T/T_0)^3$, where $k_BT_0 = \hbar\omega(N/\zeta(3))^{1/3}$ and $\zeta(3) \simeq 1.2$ is the Riemann Zeta function, while in the macroscopic limit the temperature exponent for the condensate fraction is $3/2$. The effect of the interactions in the thermal cloud is much weaker than that in the condensate, as in this case the relevant parameter is $N^{1/6}a/a_{ho}$ [66]. In a gas with repulsive interactions the interaction energy of the condensate increases significantly the total cloud energy over the ideal gas predictions (see Fig. 4 and Sec. 3.1.1).

The interactions also shift the value of the critical temperature for condensation. A mean-field negative correction of about $0.05T_0$ is predicted for a gas under harmonic confinement [68], in agreement with experiment [69]. For a homogeneous gas, instead, a shift in the critical temperature only arises in calculations beyond the mean-field level and is predicted to be positive and linear in the parameter $n^{1/3}a$ (see Sec. 3.2).

(iii) Dynamical properties. The low-lying collective modes of the confined cloud are quantized and can be excited by superimposing a time-dependent perturbation of given symmetry onto the harmonic trap. Dipolar "sloshing" modes, monopolar "breathing" modes, and quadrupolar and higher modes have been observed at various temperatures and their mode frequencies have been measured with rather high accuracy.

The dipole mode of the cloud in a harmonic well is an oscillation at the bare trap frequency. Such centre-of-mass sloshing motion is equivalent to viewing the cloud in a linearly accelerated frame and carries no information on the many-body state of the fluid (this is known as the generalized Kohn theorem [70,71]). The frequencies of the other modes are instead shifted by the interactions away from integer multiples of the trap frequency.

The measured spectra at "zero" temperature [72–74], where no thermal cloud is discernible, are in good agreement with the solution of the Landau equation of motion for a superfluid without dissipation (see Sec. 2.3.2). However, a normal cloud in the hydrodynamic regime would display a similar spectrum [75]. The study of the coherence properties of the cloud has thus been a key tool to demonstrate the presence of a condensate (see Sec. 1.3.3).

The spectra at finite temperature contain mode frequencies corresponding to excitations of both the condensate and the thermal cloud, and the damping rates increase with increasing temperature owing to collisions between the condensate and the thermal cloud [76,77]. Theoretical and numerical methods have been used to account for the observed spectra [78–81] (see Sec. 3.3.4).

A localized perturbation of the density by a laser beam can create propagating sound waves in an elongated condensate [82]. In this case the dispersion relation of the homogeneous fluid approximately holds, as the wavenumber is much larger than the inverse condensate size. Correlation functions can be estimated in a local density picture.

Condensates have also been excited beyond the linear regime. Harmonics generation [83], soliton propagation [84–86], and shock-wave formation [87] have all been observed.

1.3.3 Coherence properties of the condensate

Experiments aimed at revealing the coherence of the condensate have demonstrated its unique properties as a matter wave. Its behaviour is very different from those of a boson thermal cloud or of a fermion cloud and can lead to important applications such as the atom laser [88–92]. It has also allowed investigations of fundamental questions regarding the establishment and the decay of coherence [93–99].

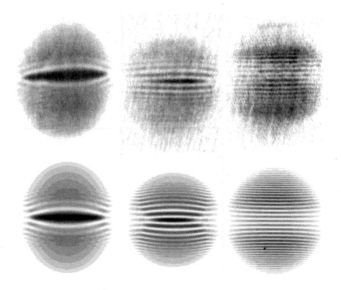

Fig. 5. Interference between two expanding ^{23}Na condensates: experiment (top) and numerical simulation (bottom). From Ketterle *et al.* [101].

(i) First-order coherence. Interference between two condensates was first observed at MIT [100] by allowing superposition of two expanding condensates that were initially trapped in a double-well potential. Interference fringes were observed in the absorption images, thus demonstrating the coherence of the condensates (see Fig. 5). Two different situations were considered: *(i)* the interference between two wholly independent condensates in the limit of a high barrier between the two wells; and *(ii)* two weakly coupled condensates in the case of lower barrier height. In the first situation the relative phase of the two condensates is different in each realization and the pattern of the fringes changes in each individual experiment, so that an average over many experiments would show no fringes at all. This is the result obtained by calculating the one-body density matrix (see *e.g.* Castin's review [102]). In the second situation instead the fringes always present a maximum at the centre of the image, reflecting the fact that the two condensates have a well determined relative phase [98,103].

Of course, coherence between two condensates is limited by thermal and quantum fluctuations, which tend to reduce the contrast between the fringes. Phase diffusion is the main mechanism responsible for decoherence in the $T = 0$ limit [99], since from Goldstone's theorem phase fluctuations cost very little energy at long wavelengths. The typical time scale for phase diffusion and disappearence of the fringes is of order $\hbar/(\sqrt{N}\partial\mu/\partial N)$.

(ii) Condensate interferometry. Condensate interferometry implies the use of a

18

condensate that has been split into two parts with a definite phase relationship between them, these parts being then brought into overlap and interference as for an optical-laser beam that has gone through a beam splitter.

Coherent splitting of a condensate has been achieved by optically induced Bragg diffraction [104]. A number of ingenious methods have been developed to extract a collimated beam of atoms from a BEC. We may mention *(i)* the application of an rf field inducing spin flips between trapped and untrapped states [90]; *(ii)* the use of optical Raman processes driving transitions between trapped and untrapped magnetic sublevels [91]; and *(iii)* the splitting of an atomic wavepacket by diffraction against an optical standing wave [105].

(iii) Higher-order coherence. Higher-order m-particle correlators are defined as

$$g_m(0) = \frac{1}{(n(0))^m} \langle (\hat{\Psi}^\dagger(0,0))^m (\hat{\Psi}(0,0))^m \rangle \qquad (18)$$

at the centre of the trap. In particular, second-order coherence is reflected in the density-density correlator. In the absence of interactions we have $g_2(0) = 2$ for a boson thermal cloud, while for a pure condensate $g_2(0) = 1$. Access to the second-order coherence function $g_2(r)$ is allowed for a contact potential by measurements of the mean interaction energy of the gas,

$$E_{int} = \frac{g}{2} \int d^3r \, g_2(r) n^2(r). \qquad (19)$$

Within experimental resolution it is found that in a dilute gas $g_2(0) = 2$ for the thermal cloud and $g_2(0) = 1$ for the condensate [106–108].

Third-order coherence gives information about three-body processes in the gas. For a non-interacting thermal cloud $g_3(0) = 3!$ while for a pure condensate we have again $g_3(0) = 1$. This means that a thermal cloud is expected to decay from three-body recombinations six times faster than a condensate, as has been measured by Burt *et al.* [109].

The measurement of the decay rate of the cloud was proposed by Kagan and coworkers [110] (see also [94]) as a method to detect the presence of a condensate and has been used to infer quantum degeneracy in a quasi-2D gas of spin-polarized hydrogen [111]. Extension of the theory to the 2D case was done by Kagan *et al.* [112], who introduced the notion of a quasicondensate. Simulation of a gas of ultracold atoms in a quasicondensate has been carried out [113] using the quantum Worm algorithm (see Sec. 3.2.1).

1.3.4 Superfluidity

Several aspects of superfluidity have been probed in confined atomic gases. The analogue of the response to a transverse probe is given by the excitation

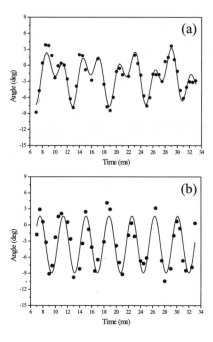

Fig. 6. Experimental results for the scissors modes of a thermal cloud (a) and of a condensate (b). From Maragó et al. [114].

of the scissors modes, which are peculiar to finite superfluid systems and were previously known in nuclear matter. Quantization of vortex lines has been demonstrated and the breakdown of superfluidity has been studied in a series of experiments.

(i) Scissors modes. Scissors modes are excited in an atomic gas by a sudden twist of its anisotropic trap [114]. This perturbation gives rise to oscillations at a single frequency in the case of a condensed cloud, but showing beats between two frequencies for a non-condensed cloud [115] (see Fig. 6). While only longitudinal modes can be excited in a superfluid, the applied perturbation excites both longitudinal and transverse modes in a thermal cloud.

(ii) Quantized vortices. Vortices have been created in atomic condensates with a variety of techniques: by stirring the condensate with a laser beam [116], by exploiting interconversion between two components of the condensate with different spins [117], or by using topological phases [118]. Above a critical stirring frequency first one and then several vortices are observed (see Fig. 7), which ultimately arrange themselves into a triangular lattice [120].

A peculiarity of superfluids is vortex quantization (see Eq. (13)). Quantization of a single vortex line has been demonstrated experimentally using the idea

20

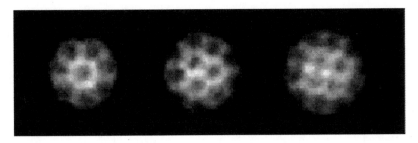

Fig. 7. Experimental observation of vortices in a BEC of ^{87}Rb atoms. From Chevy *et al.* [119].

that the presence of a vortex line inside the condensate breaks time-inversion symmetry and leads to a splitting of the quadrupolar surface modes of azimuthal quantum numbers $m = \pm 2$. As a consequence, a slow precession is induced in a quadrupolar distortion of a condensate containing a vortex at its centre and the frequency splitting is proportional to the quanta of vortex circulation (see *e.g.* the review of Fetter and Svidzinsky [121]). This method has been used to measure single quanta of circulation at ENS [122] and multiple quanta at MIT [118]. As an alternative, an interferometric technique has been used to map the phase profile of the condensate wavefunction in a path around the vortex [123].

(iii) for breakdown of superfluidity. In early experiments on liquid ^{4}He it was found that superfluid flow persists only at low velocities and dissipation sets in above a threshold. The Landau criterion for breakdown of superfluidity [26] states that dissipation starts during flow when the creation of elementary excitations becomes energetically favourable in the moving fluid, leading to their spontaneous emission. The critical velocity is the speed of sound if a roton minimum is absent in the dispersion curve of the excitation spectrum.

The threshold for the breakdown of superfluidity in confined gases has been investigated in various experiments, probing the condensate on both a macroscopic and a microscopic scale. Onofrio *et al.* [124] have studied hydrodynamic flow in a condensate stirred by a blue-detuned laser beam acting as a macroscopic moving object. A density-dependent critical velocity for the onset of a distortion in the density distribution was observed, the distortion being associated with a pressure gradient from the drag force between the beam and the condensate. The measured critical velocity is considerably lower than the local speed of sound and arises from the periodic shedding of vortices at a rate that increases with velocity.

Microscopic probes colliding with a BEC have been realized by adding moving impurities [125]. A drastic reduction of the collision rate is measured when the velocity of the impurity is reduced below the sound velocity, thus providing an experimental test of the Landau criterion.

21

Breakdown of superfluidity and onset of decoherence have also been observed in a condensate placed in a magnetic trap and driven through a weak 1D optical lattice [126]. The dynamical evolution of the condensate was controlled by displacing the trap by an amount Δx in the direction parallel to the lattice. Different dynamical regimes are observed depending on the magnitude of Δx: for small displacements the condensate manifests its superfluid behaviour by performing undamped oscillations in the harmonic well, but at larger values of the displacement, which correspond to a higher value for the maximum velocity attained by the fluid during its motion, the onset of dissipative processes generates a diffusing cloud of non-condensed atoms. The observations in the dissipative regime can be quantitatively interpreted in terms of a density-dependent local critical velocity for the destruction of superfluidity, given by the local speed of sound $v_c(x) \propto \sqrt{n(x)}$. This kind of instability is different from the one that is observed for higher lattice barriers, which is attributed to the existence of an unstable branch of Bogoliubov excitations in the lattice [127].

1.3.5 Condensates in optical lattices

A 1D optical lattice as mentioned just above is created for an atomic gas as a standing wave from two counterpropagating and blue-detuned laser beams. The resulting dipole-force potential felt by the atoms is proportional to the laser intensity and has the form

$$U(x) = U_0 \cos^2(k_L x), \tag{20}$$

where k_L is the laser wavenumber. Lattices in higher dimensions and with different geometries are obtained by combining pairs of standing waves along various directions with suitable polarizations of the laser beams.

Early experiments on cold thermal atoms in an optical lattice have demonstrated solid-state phenomena such as Bloch oscillations and Landau–Zener tunnelling [128, 129], the effect of a constant force applied to the atoms being analogous to an electric field acting on charge carriers in a semiconductor. The peculiarity of the condensate is its phase coherence and the possibility of attaining strongly correlated states.

(i) Ground state and excitations. A condensate that has been adiabatically loaded into an optical lattice keeps phase coherence between its fragments in different sites, provided that the lattice barrier is not too high and still allows tunnel between neighbouring sites. Phase coherence and periodic occupation of a 2D square lattice have been demonstrated in expansion experiments [130]: the time-of-flight images, which reflect the momentum distribution of the gas, show an interference pattern with its maxima on reciprocal-lattice sites, with

separation $2\hbar k_L$ in each direction of momentum space as fixed by the inverse of the lattice period $d = \pi/k_L$.

An infinitely extended condensate at rest in an optical lattice sits in the $\mathbf{q} = 0$ quasi-momentum state at the chemical potential. The theory of elementary excitations of such a condensate involves concepts which are well known in solid state physics. Considering here for simplicity the case of a 1D lattice, the energy-momentum dispersion relation is a multi-valued periodic function forming the "energy bands" $E_n(q)$ inside the Brillouin zone $(-k_L < q \leq k_L)$. The corresponding single-particle states obey Bloch's theorem and gaplessness is ensured by constructing these states from two sets of Bloch orbitals having definite symmetry under time reversal [131, 132], namely $Z_{nq}^{\pm}(x) = 2^{-1/2}[u_{nq}(x) \pm v_{nq}(x)]$ where the functions $u_{nq}(x)$ and $v_{nq}(x)$ are the Bogoliubov amplitudes and obey the Bogoliubov-de Gennes equations. A dispersion relation starting out linearly from the zone centre for the elementary excitations in the lowest positive-energy band follows from the gaplessness property and is again a consequence of the Goldstone theorem. The use of the GPE in calculating the Bloch states in this band yields instead a quadratic dispersion near the zone centre [133, 134] (see Sec. 2.3.4). The possibility that nonlinearity may introduce loop structures at special points of the dispersion relation has been discussed by several authors [135-137].

It is convenient to express the Bloch orbitals as superpositions of Wannier functions $w_{n\pm}(x)$ centred on the lattice sites, namely

$$Z_{nq}^{\pm}(x) = N^{-1/2} \sum_l \exp(iqld)w_{n\pm}(x - ld) \qquad (21)$$

where the index l runs over all N sites. Only one Wannier function is associated with each energy band in 1D [138]. It is then easy to show [132] that (i) the whole of the lowest band corresponds to pure modulations of the phase of the condensate, and (ii) the higher bands derive from the superposition of localized excitations with a definite phase relationship. This formulation reduces to a tight-binding picture as the height of the lattice potential barrier is increased.

In practice a real condensate in an optical lattice has finite length L_i along each space direction, which may correspond to the occupation of up to a few hundred lattice sites . The indeterminacy in the quasi-momentum is $\Delta q_i \sim 2\pi/L_i$ and thus is a small fraction of the width of the Brillouin zone.

(ii) Bragg scattering. Pulsed optical lattices have been used for the study of the microscopic excitations of the Bose fluid. The dynamic structure factor and the dispersion relation of the excitations can be accessed by applying for some time two slightly detuned laser beams with energy difference δ and relative angle θ, which produce resonant excitations with momentum $q = 2k_L \sin(\theta/2)$ and energy δ [139-141].

In the regime of free-particle-like dispersion one can access the momentum distribution of the BEC, as in the case of the hydrogen condensate where direct imaging could not be used [58]. In a quasi-1D condensate the momentum distribution, being the Fourier transform of the one-body density matrix, allows one to measure the length over which the condensate loses phase coherence from fluctuations [142]. A further application of repeated Bragg pulses has been the demonstration of phase coherence in a 3D trapped condensate all along its size [143].

(iii) Coherent tunnelling. The coherent motions of a condensate through an optical lattice have been studied by driving it with external forces. In particular, a condensate subject to gravity in a lattice emits a sequence of coherent drops [89] with period $T_B = h/(mgd)$ of Bloch oscillations as determined by the acceleration of gravity, the atomic mass, and the lattice spacing. Such drop emission is due to the simultaneous spill-out of atoms from various lattice sites, in analogy with Landau-Zener tunnelling into the continuum.

Band structure for a condensate in a lattice may be controlled by suitable arrangements of laser beams and optical mirrors. In particular, interference between Bragg scattered and inter-subband tunnelling wave packets is predicted to occur when the periodicity of the lattice is doubled and to be revealed through drop emission under a constant external force [144]. Quasi-periodic structures may be also created by the same basic optical tools.

The experiment of Burger *et al.* [126] already presented in Sec. 1.3.4 involves driving a condensate through a quasi-1D optical lattice with a harmonic force. In the superfluid regime corresponding to small displacements of the harmonic bowl, the behaviour of the condensate can be mapped into that of a superfluid current flowing through a weak-link Josephson junction under an applied a.c. voltage [145] (see also [146] and Sec. 2.3.5).

(iv) Strongly correlated states. On increasing the lattice barriers and decreasing the atomic population at each lattice site the gas is driven towards a strongly correlated regime, where the interaction energy starts to dominate. Phase coherence through the lattice is being lost in this regime and, since phase and particle number at each site are canonically conjugate variables, number fluctuations decrease as phase fluctuations increase. Squeezed states with reduced number fluctuations have been observed by Orzel *et al.* [147].

On further increase of the lattice barriers the gas enters the Mott-insulator regime, where the atom-atom repulsions induce a commensurate filling of the lattice sites and a gap in the excitation spectrum [148]. The Mott insulator phase has been observed in a 3D lattice of ^{87}Rb atoms [149], where the disappearance of phase coherence is revealed by the replacement of interference

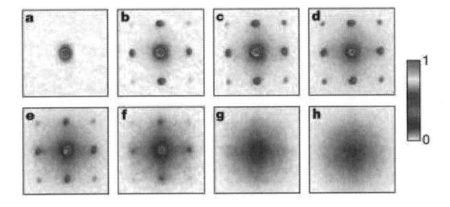

Fig. 8. Transition to the Mott-insulator state as observed for a BEC in a lattice through the blurring of the interference pattern as the barrier height is increased. From Greiner *et al.* [149].

dots in the expansion image by a broad distribution (see Fig. 8). The transition to the Mott-insulator phase has been the object of extensive numerical studies (see Sec. 2.2.5). This state has been suggested as a possible "register" in quantum computation [150], which would allow gate operations by exploiting atom-atom collisions [151]. Later experiments have demonstrated that one every three sites of the lattice can be loaded [152] and that a two-qubit gate can be operated with atoms in an optical lattice [153].

Exotic phases and strongly correlated states can also be achieved by trapping spinorial condensates [154] or dipolar gases [155] in a lattice.

1.4 Confined Fermi gases and boson-fermion mixtures

Similar techniques to those employed to trap dilute gases of bosonic atoms can be applied to gases of fermionic isotopes. The quantum statistics of the fermions makes the cooling process more challenging. New aspects of the physics of quantum fluids can be explored, such as the shell structure in the density distribution, the suppression of collision rates, changes in excitation spectra and in damping rates, and fermion pairing and superfluidity. Mixtures of bosons and fermions with a tunable coupling strength allow one to explore new phases, such as demixing of the two species and boson-induced fermion pairing, as well as instabilities such as the collapse of the fermion cloud.

1.4.1 Cooling fermions

^{6}Li and ^{40}K are the only stable fermionic species among the alkali atoms. Whereas the standard evaporative cooling scheme is based on s-wave collisions as this is the only effective channel at ultra-low temperatures below $100\mu K$, identical fermions in a magnetic trap cannot collide in this channel owing to the antisymmetry of the fermionic wavefunction. The cooling problem has been tackled by trapping either a fermion gas of two components in different internal states or two different atomic species. The interspecies s-wave collisions are allowed and "symphatetic" evaporative cooling can be implemented.

Quantum statistics still influences the evaporative cooling trajectory as the gas is cooled below the Fermi temperature T_F, given by $k_B T_F = \hbar\omega(6N_F)^{1/3}$ for N_F non-interacting fermions under isotropic harmonic confinement with frequency ω. The evaporation of a two-component Fermi gas becomes inefficient when the temperature is lowered to below about $0.5T_F$ [156], as the states near the Fermi level are increasingly populated so that less and less final states are available for collisions. Use of a boson-fermion mixture reduces this "Pauli blocking" effect and also increases the accuracy of temperature measurements, but becomes inefficient below $0.3T_F$ when the bosonic specific heat becomes lower than the fermionic one [157]. Another limitation to fermion cooling in these mixtures can be boson superfluidity: a fermion is not scattered by a BEC if its incoming velocity is lower than the bosonic speed of sound [158].

A motivation for further fermion cooling has been the search for novel superfluid states. Early theoretical estimates [159] suggested that a BCS-like state made from s-wave pairing of fermions in two different internal states could be found below $0.01T_F$ (see Sec. 1.4.4). Cooling of fermions down to $0.05T_F$ has been achieved in boson-fermion mixtures by ejecting both species from the trap ("dual cooling"), thus keeping a substantial thermal fraction of bosons in the cloud [160], or by following an out-of-equilibrium evaporation path [161].

At the time of writing ultracold Fermi gases have become available for experiments in the following mixtures: ^{6}Li - ^{6}Li [156,162–164], ^{6}Li - ^{7}Li [157,165], ^{40}K - ^{40}K [166], ^{40}K - ^{87}Rb [167,168], and ^{6}Li - ^{23}Na [169]. An example of a boson-fermion mixture is reported in Fig. 9.

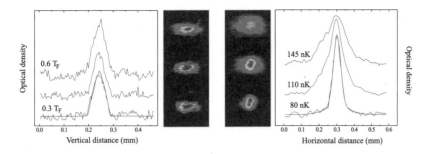

Fig. 9. A mixture of ^{87}Rb gas and ^{40}K gas from the experiment of Roati *et al.* [167]: optical density of ^{40}K fermions (left) and of ^{87}Rb bosons (right) at three values of the temperature.

1.4.2 Effects of quantum degeneracy

A number of experiments on confined Fermi gases have highlighted two main aspects of Fermi statistics. These are the so-called Fermi pressure and the relaxation behaviour of the oscillations of a fermion gas.

(i) Fermi pressure. At zero temperature the fermions fill the energy levels of the harmonic trap up to a maximum fixed by their number N_F. The role of the Pauli principle in keeping the spin-polarized atoms apart is referred to as the Fermi pressure, which acts as an effective repulsion even between non-interacting fermions. The Fermi pressure is believed to stabilize white dwarfs and neutron stars against gravitational collapse [170].

The saturation of the mean square radius of a fermion cloud at low temperature has been one of the early observations of the consequences of Fermi statistics [165,157]. A similar effect of Fermi pressure is reflected in the total energy of the gas: considering for simplicity a non-interacting gas, at decreasing temperature its energy deviates from the classical value $3N_F k_B T$ and saturates towards the value $(3/4)N_F k_B T_F$ at $T = 0$. Also this property has been experimentally verified [156].

For small atomic samples or elongated geometries the shell effects arising in the density profile from the single occupation of the levels should become observable (see Sec. 4.1).

(ii) Spin excitations and thermal relaxation. In a fermion mixture it is possible to study the collective modes of the two spin components [171,172]. For ^{40}K atoms in a cigar-shaped trap both the collisionless and the hydrodynamic regime have been reached in the excitation of dipolar modes, the collisionality parameter in the gas being varied by changing the total number of particles and the radial confinement strength.

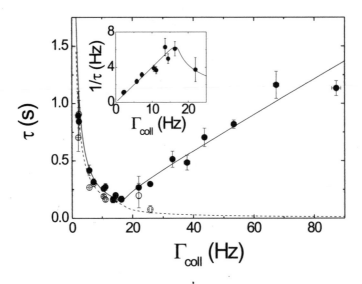

Fig. 10. Transition from collisionless to hydrodynamic behaviour in a two-component gas of ^{40}K atoms from the experiment of Gensemer and Jin [166].

Experiments performed on such two-component fermion clouds [166,173] have shown how the frequencies of the two spin dipole modes tend to lock at the same intermediate value as the collisionality is increased. The damping time of the oscillations goes through a minimum as the gas is driven from the collisionless to the hydrodynamic regime (see Fig. 10). The quantum fluid displays well-defined collective modes in both regimes, the damping being stronger in the intermediate region of collisional rates. A particle dynamics simulation has explored this dynamical transition down to the quantum degenerate regime (see Sec. 4.2).

Pauli blocking also affects the thermal relaxation rate: when one of the two fermionic components is driven out of equilibrium by suddenly decreasing its particle number, the relaxation time towards its new equilibrium size gets longer and longer as the temperature is decreased [174].

1.4.3 Instabilities in boson-fermion mixtures

The interactions between bosons and fermions can lead to dramatic changes in the state of a boson-fermion mixture, *i.e.* implosion for attractive coupling or spatial demixing for repulsive coupling.

(i) Collapse. A ^{87}Rb-^{40}K mixture is characterized by large attractive interactions between its components, the measured value of the boson-fermion scattering length being $a_{BF} = -21.7 \pm 4$ nm. The interaction energy is determined by the overlap of the two density profiles, so that the strength of the boson-fermion coupling can be tuned by varying the number of bosons and can be made large enough to overcome the Fermi pressure and induce collapse of the cloud. The collapse of the fermionic component has been observed as a sudden disappearance of the fermion cloud from the trap [175]. Indeed, the transition depends so sharply on a_{BF} that the measurement of the collapse point can be used to determine its value [176].

In the experiment the increase of the particle densities at collapse is halted by three-body (K-Rb-Rb) recombinations. As these become an important loss term, the number of bosons in the trap is reduced back below the threshold for collapse.

(ii) Demixing. The equilibrium state of a mixture with repulsive boson-fermion and boson-boson interactions depends on the relative strength of the various couplings. If the boson-fermion coupling is stronger than the boson-boson one and can overcome the effective repulsion arising from the Fermi pressure, then the two components of the mixture will tend to occupy distinct regions in space.

Spatial demixing has not yet been observed in boson-fermion mixtures at time of writing, although an experiment on ^{6}Li-^{7}Li [165] seems to be quite close to the demixing condition [177]. Sec. 4.3 discusses how to locate the transition in a mesoscopic cloud and the possible configurations of the cloud in the demixed state.

1.4.4   Fermion superfluidity

Attractive interactions between fermions may lead to a paired state with superfluid properties. In parallel with experimental efforts to achieve superfluidity in gases of fermionic atoms, the theory has been providing suggestions and possible strategies to reach, detect, and characterize such a superfluid state.

Several mechanisms have been envisaged for fermion pairing. In the dilute regime a BCS-like state is predicted, with the formation of large-sized Cooper pairs. However, in a spin-polarized Fermi gas the interactions occur at best in the p-wave channel and, if they are attractive, the critical temperature in the homogeneous state is predicted to be $T_c \sim T_F \exp(-\pi/[2(k_F|a_p|)^3])$ where a_p is the p-wave scattering length [178]. This is far too low to be reached at present. The critical temperature might be increased in the presence of another species, either fermionic [179] or bosonic [180], since phonon fluctuations can mediate further effective attractions.

Another possibility for a BCS-like state is a two-component Fermi gas with attractive interactions that may lead to the formation of Cooper pairs in the s-wave channel. The critical temperature in this case would be $T_c \sim T_F \exp(-\pi/[2(k_F|a_s|)])$, where a_s is the intercomponent scattering length [159]. For ^{6}Li a_s is very large and negative, suggesting it as a possible candidate for achieving BCS-like superfluidity. The critical temperature is highest when the populations of the two spin states are the same and the addition of a bosonic component to the mixture might again increase the value of T_c [181,182].

The value of the scattering length can be varied by some orders of magnitude by exploiting Feshbach resonances (see Sec. 1.4.5) and one may in this way reach a strong-coupling regime where the size of the Cooper pairs is much smaller than their mean separation [183]. In such limit the critical temperature for superfluidity is approximately equal to that for condensation of bosons with twice the fermion mass and half the fermion density. This leads to $T_c \sim 0.2 T_F$ [184,185] (see also Sec. 4.4.1).

In the BCS limit the density profile and the momentum distribution of the gas are only weakly affected by the pairing transition. Several alternative observations have been suggested for detecting the transition: a Raman transition analogous to tunnelling across a normal-superconductor junction [186], the spectrum of collective modes including spin excitations and scissors modes [187–189], the moment of inertia [190], Bragg scattering of light across the gas [191], and an anisotropic expansion after release from the trap [192]. In the BEC limit the condensate of Cooper pairs has been predicted to already be visible as a peak in the density profile [193].

1.4.5   Towards strongly correlated fermions

The effective strength of the interactions in dilute atomic gases can be modified in several ways. In addition to varying the density of the sample (see Sec. 1.4.2) or to localizing the atoms on a lattice (see Sec. 1.3.5), one may exploit a Feshbach resonance. This occurs when the collision energy of two free atoms is the same as that of a quasi-bound molecular state and on approaching it the scattering length can increase by some orders of magitude and also change sign [194–200]. The dilution condition breaks down as the gas is driven close to a Feshbach resonance.

Feshbach resonances have been used to create a stable condensate of ^{85}Rb atoms [59] and to generate bright solitons in ^{7}Li [201,202]. A large enhancement of losses is usually observed near the resonance and for a condensate of ^{87}Rb atoms has been related to the formation of diatomic molecules *via* three-body processes. A coherent mixture of atomic and molecular states has been created in ^{85}Rb and probed by sudden changes in magnetic field, which

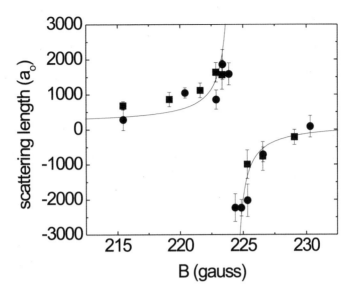

Fig. 11. Scattering length for ^{40}K atoms in two different Zeeman sublevels driven across a Feshbach resonance by varying the magnetic field B, as measured by Regal and Jin [208].

lead to oscillations in the number of atoms remaining in the condensate [203]. The prospect of creating a superposition of atoms and molecules has been considered in a number of theoretical papers (see *e.g.* [204] and references therein).

Feshbach resonances have also been studied in fermionic alkali gases [163, 164, 205-207] (see Fig. 11) and the presence of molecules near the resonance has been inferred from the data [160, 209, 210]. Contrary to the bosonic case, molecules formed with fermionic atoms are extremely stable and could be cooled down to degeneracy or even brought into a superfluid state [211].

In experiments carried out on spin mixtures of ^{40}K atoms by Greiner *et al.* [212] and of ^{6}Li atoms by Jochim *et al.* [213] and by Zwierlein *et al.* [214], formation of diatomic molecules has been realized by evaporative cooling in an optical dipole trap near a Feshbach resonance. Further cooling of the molecular gas has led to the observation of a Bose-Einstein condensate. The condensate is revealed by the emergence of a peak in the density distribution (see Fig. 12), as predicted for a condensate of Cooper pairs [193].

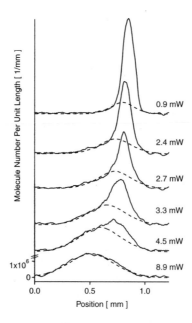

Fig. 12. Onset of BEC in a gas of $^6Li_2$ molecules. A bimodal distribution appears in the radial average of the absorption images as the temperature is lowered (from bottom to top). From Zwierlein *et al.* [214].

Two further examples of phenomena that are met in gases of strongly correlated fermions may be mentioned in closing. Studies of the dynamics of a highly collisional gas can shed light on novel aspects of hydrodynamic behaviour. In particular, a consequence of high collisionality is an anisotropic expansion of a fermion cloud after release from the trap [208, 215] and this is not necessarily a signature of superfluidity [216]. Secondly, under some conditions a strongly coupled gas cannot be described by mean-field equations. For example, a two-component Fermi cloud is observed to be stable under very strong attractive interactions, in conditions such that the mean-field equations predict a collapse [217]. According to many-body theory the behaviour of the gas for very large values of the scattering length a is expected to be independent of a and determined only by the particle density [218]. It would be interesting to test such "universal" behaviour on experimental observables.

2 Bose gases at zero temperature

In this Chapter we review work on quantum degenerate Bose gases under confinement in conditions such that thermal excitation can be neglected. In the high-dilution limit, where the gas is pratically all Bose-Einstein condensed, the ground state is well described by the mean-field Gross-Pitaevskii equation (GPE). In broad terms the GPE is a three-dimensional nonlinear Schrödinger equation, for which a variety of consolidated numerical techniques are available. This equation has been used to quantitatively understand the results of several experiments on condensed alkali gases, to predict scaling laws, and to study the optical properties of a condensate behaving as a matter wave.

In atomic gases one may reach strongly correlated regimes as the scattering length is increased, or the dimensionality is reduced by modifying the external potential (*e.g.* by using tight harmonic traps or optical lattices), or disorder is added. For a proper description of these situations one needs to resort to a many-body Schrödinger equation, which is a linear partial differential equation in a space of dimensionality $3N$. Due to this hyper-dimensionality this equation can usually be solved only by a specific class of stochastic techniques, namely by the Quantum Monte Carlo method. Non-trivial ground states and new phases can then emerge.

The excitation spectrum of a degenerate Bose gas reflects its superfluid properties and possible quantum phase transitions. Again, the time-dependent version of the GPE accounts for the dynamical behaviour of the gas in the dilute limit, but fails to describe situations in which a significant quantum depletion of the condensate is induced by the interactions, by fluctuations, or by external drives.

Here we aim to give an overview of the main techniques which have appeared so far in the BEC literature for both types of equations and to illustrate their use by reporting some physical applications. For ground-state calculations based on the GPE these techniques include variational methods, boundary eigenvalue solvers, as well as various methods to advance the GPE in imaginary time mostly through finite-difference discretizations. This presentation will be followed by a Section dealing with work on many-body effects, with main attention to the usefulness of the Diffusion Monte Carlo method in the study of Bose gases. We then turn to the main numerical techniques to solve the time-dependent GPE. These include various forms of grid discretization, both in configuration and in reciprocal space, followed by either explicit or implicit time marching. Due to the enormous wealth of the literature on the subject, a fully exhaustive coverage is virtually impossible and we apologize in advance for inadvertent omissions. Our main task will be to give some major guidelines to the reader, leaving the details to a study of the original papers.

The basic idea of a mean-field description of a dilute Bose gas dates back to Bogoliubov (see Sec. 1.2.2) and is easily extended to inhomogeneus and time-dependent configurations. The bosonic field operator $\hat{\Psi}(\mathbf{x}, t)$ is expressed as the sum of a "slow" background field plus a fluctuating perturbation,

$$\hat{\Psi}(\mathbf{x}, t) = \Phi(\mathbf{x}, t)\hat{a}_0 + \tilde{\Psi}(\mathbf{x}, t) . \tag{22}$$

Here $\hat{a}_0$ is the annihilation operator for particles in the condensate, and $\Phi(\mathbf{x}, t) \equiv \langle \hat{\Psi}(\mathbf{x}, t) \rangle$ is a complex function defined as the expectation value of the field operator on a suitably defined ensemble (see Sec. 1.2.4) and describes a classical field playing the role of an order parameter. The amplitude of this function determines the condensate density through $n_0(\mathbf{x}, t) = |\Phi(\mathbf{x}, t)|^2$. It can be shown that $\Phi(\mathbf{x}, t)$ is the eigenvector associated to the largest eigenvalue of the one-body density matrix $\rho(\mathbf{x}', \mathbf{x}; t) = \langle \hat{\Psi}^\dagger(\mathbf{x}', t)\hat{\Psi}(\mathbf{x}, t) \rangle$ ([21,219]).

The decomposition (22) is especially useful when the fluctuactions of the condensate, which are described by $\tilde{\Psi}(\mathbf{x}, t)$, are small. An equation for the order parameter can then be derived by expanding the theory to the lowest order in $\tilde{\Psi}$, as in the case of the uniform Bose gas. The crucial difference from the homogeneus case is that already the "zero-th order" theory for $\Phi(\mathbf{x}, t)$ delivers a non-trivial dynamics.

The equation for the condensate wavefunction $\overset{\circ}{\Phi}(\mathbf{x}, t)$ is derived by starting from the second-quantized Hamiltonian,

$$\hat{\mathcal{H}} = \int d\mathbf{x}\, \hat{\Psi}^\dagger(\mathbf{x}) \left[-\frac{\hbar^2}{2m}\nabla^2 + V_{\text{ext}}(\mathbf{x}) \right] \hat{\Psi}(\mathbf{x})$$
$$+ \frac{1}{2} \int\int d\mathbf{x} d\mathbf{x}'\, \hat{\Psi}^\dagger(\mathbf{x})\hat{\Psi}^\dagger(\mathbf{x}')V(\mathbf{x} - \mathbf{x}')\hat{\Psi}(\mathbf{x}')\hat{\Psi}(\mathbf{x}). \tag{23}$$

Here $V_{\text{ext}}(\mathbf{x})$ is the externally applied potential and $V(\mathbf{x} - \mathbf{x}')$ is the two-body interatomic potential. The evolution equation for the operator $\hat{\Psi}(\mathbf{x}, t)$ then is

$$i\hbar\, \partial_t\, \hat{\Psi}(\mathbf{x}, t) = \tag{24}$$
$$\left[-\frac{\hbar^2\nabla^2}{2m} + V_{\text{ext}}(\mathbf{x}) + \int d\mathbf{x}'\, \hat{\Psi}^\dagger(\mathbf{x}', t)V(\mathbf{x} - \mathbf{x}')\hat{\Psi}(\mathbf{x}', t) \right] \hat{\Psi}(\mathbf{x}, t) .$$

For a dilute gas the operator $\hat{\Psi}$ can be replaced by the classical field Φ. As discussed in Sec. 1.2.1, it is customary to choose the interatomic potential in the form of an effective point-like interaction,

$$V(\mathbf{x} - \mathbf{x}') = g\delta(\mathbf{x} - \mathbf{x}') \tag{25}$$

where the coupling constant g is related to the s-wave scattering length a by $g = 4\pi\hbar^2 a/m$. The use of the effective interaction (25) yields the following closed equation for the order parameter,

$$i\hbar\partial_t\Phi(\mathbf{x},t) = \left(-\frac{\hbar^2\nabla^2}{2m} + V_{\text{ext}}(\mathbf{x}) + g|\Phi(\mathbf{x},t)|^2\right)\Phi(\mathbf{x},t) . \qquad (26)$$

This is the GPE, which was derived independently by Gross [220,221] and by Pitaevskii [222]. Its validity rests on the condition that the scattering length be much smaller than the mean distance between the atoms and that the number of atoms in the condensate be large. The GPE can be used to explore the behaviour of the gas at very low temperature for variations of the order parameter that occur over distances larger than the mean interatomic distance.

A condensate at equilibrium lies at the chemical potential μ (see $e.g.$ [12]). By setting $\Phi(\mathbf{x},t) = \Phi(\mathbf{x})\exp(-i\mu t/\hbar)$ in Eq. (26) one obtains the static GPE,

$$\left(-\frac{\hbar^2\nabla^2}{2m} + V_{\text{ext}}(\mathbf{x}) + g|\Phi(\mathbf{x})|^2\right)\Phi(\mathbf{x}) = \mu\Phi(\mathbf{x}) . \qquad (27)$$

Equation (27) can also be derived by variational minimization of the energy functional

$$E[\Phi] = \int d\mathbf{x} \left[\frac{\hbar^2}{2m}|\nabla\Phi|^2 + V_{\text{ext}}(\mathbf{x})|\Phi|^2 + \frac{g}{2}|\Phi|^4\right] . \qquad (28)$$

The three terms in the integral are the kinetic energy E_{kin} of the condensate, the external potential energy E_{ext}, and the mean-field interaction energy E_{int}. For repulsive interactions ($g > 0$) the functional is convex and its minimum delivers the stable ground-state of the gas. For $g < 0$ the ground state exists only at low coupling for limited numbers of bosons in the trap, as long as the zero-point energy balances the effect of attractions and prevents collapse (see Sec. 2.3.8). A similar density functional approach has been proposed for low-dimensional Bose gases [223], where the interaction energy has a different dependence on density ($\log|\Phi|^2$ in 2D and $|\Phi|^6$ in 1D in the Tonks gas regime, see Sec. 2.3.9).

The early experiments were performed on BEC's inside magnetic traps with harmonic confinement (see Sec. 1.3.1). The ground state of these confined condensates can be computed by numerical solution of the GPE. In the ideal gas the ground-state wavefunction is

$$\Phi(r) = N^{1/2}(2\pi a_{ho})^{-3/2}e^{-r^2/(2a_{ho}^2)} \qquad (29)$$

in the isotropic case $V_{ext}(r) = m\omega^2 r^2/2$, with $a_{ho} = (\hbar/m\omega)^{1/2}$ and chemical potential $\mu = 3\hbar\omega/2$. In the opposite limit ($E_{int} \gg E_{kin}$) a useful approxima-

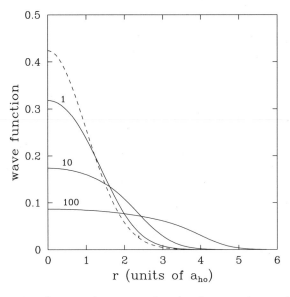

Fig. 13. Illustrating the ground-state wavefunction for a non-interacting Bose gas (dashed line) and for increasing atom-atom repulsions. Redrawn from Dalfovo *et al.* [2].

tion is provided by the Thomas-Fermi theory,

$$
\begin{cases}
\Phi(r) = \sqrt{(\mu - V_{ext}(r))/g} & \text{if} \quad V_{ext}(r) < \mu \\
\Phi(r) = 0 & \text{if} \quad V_{ext}(r) > \mu
\end{cases}
$$

with $\mu = (15Na/a_{ho})^{2/5}\hbar\omega/2$. This value is obtained by fixing the number N of bosons in the condensate with the condition $N = \int d\mathbf{r}\, |\Phi(r)|^2$.

Figure 13 reports an example of the ground-state wavefunction of a BEC for an ideal gas and for several values of the repulsive coupling strength. The atom-atom repulsions considerably broaden the density profile while the kinetic energy determines its tails, which reflect the spill-out of the atoms by tunnel into the walls of the trap. The Thomas-Fermi approximation cuts these tails off at $R = \sqrt{2\mu/m\omega^2}$ and this requires large numbers of atoms ($N > 5000$, say). An accurate numerical solution of the static GPE is necessary to obtain the initial conditions for calculations of the dynamics of the condensate.

2.1.1 Variational methods

The ground-state ψ_0 of a quantum system obeys a variational minimum principle on the energy functional,

$$E[\psi] \geq E[\psi_0] \equiv E_0 \tag{30}$$

where ψ is a generic wavefunction with the appropriate symmetry. If the energy functional is known, one can construct a test function $\psi_T(\mathbf{x}; p)$ where p denotes a set of free parameters, compute the energy functional as a function of p by means of any standard numerical technique, and locate the value p^* at which $E(p)$ attains its minimum E^*. Of course, there is no guarantee that E^* closely agrees with the true ground-state energy E_0.

In their early work Baym and Pethick [224] have given a variational estimate of the ground-state energy of a BEC described by the GPE, by adopting as their variational Ansatz a Gaussian wavefunction with adjustable width. With reference to the newly realized condensate of ^{87}Rb Baym and Pethick showed that the kinetic energy becomes negligible as the number of atoms in the cloud becomes large, and that the spatial structure and the momentum distribution of the cloud depend in an essential way on the interactions. They also estimated the superfluid coherence length and the critical angular velocity for vortex stabilization.

As a further example of a variational calculation we may cite a study of a condensate with attractive interactions in a double-well potential, where the variational parameters are the fractional occupations of gerade and ungerade wavefunctions [225]. It was shown in the macroscopic limit by Nozières and Saint James [226] (see also Baym [227]) that condensate fragmentation would cost an extensive amount of exchange energy in the case of repulsive interactions, but would be favoured for attractive interactions.

A rather general strategy to solve variational problems can be based on the knowledge of the ground state in the opposite limits of weak coupling and strong coupling. One may then adopt a linear combination of the two limiting wavefunctions to form the basis for a useful variational approach. This method has been used *e.g.* for 1D Bose gases [228].

2.1.2 Numerical eigenvalue solvers

Ground-state calculations for quantum systems are often pursued by numerical eigensolvers providing the lowest-lying eigenvalue of the Hamiltonian along with the corresponding eigenfunction. The nonlinear character of the GPE implies that a whole sequence of linear eigenvalue problems has to be solved before the true ground state is reached. Such a sequence reads generically as

37

follows,

$$\mathcal{H}_{GP}(\Phi^{(k-1)})\Phi^{(k)} = \mu^{(k)}\Phi^{(k)}, \tag{31}$$

where $k = (0, \ldots, k_{max})$ and $\mathcal{H}_{GP}$ is the Gross-Pitaevskii operator. Many standard methods are available to solve a sequence of linear eigenvalue problems, such as inverse iteration and Lanczos methods [229].

Some studies in the BEC literature are moving in the direction of solving the nonlinear eigenvalue problem through systematic variational techniques. For example Bao et $al.$ [230] have computed the BEC ground state in effectively 1D configurations by direct minimization of the energy functional (28), using a finite-element representation of the wavefunction. For the spherical 3D case one has $\Phi(r) \sim \Phi_M(r) = \sum_j \phi_j\, e_j(r)$ where the M amplitudes ϕ_j are the variational parameters and $e_j(r)$ are standard piecewise-linear elements. The corresponding discretized energy functional reads

$$F_M = E_M(\phi_1, \ldots, \phi_M) - \frac{\lambda}{2}\left(\int |\sum_{j=1}^{M} \phi_j e_j(r)|^2\, 4\pi r^2 dr - 1\right) \tag{32}$$

where the second term on the r.h.s. introduces a Lagrangian multiplier enforcing normalization. The minimum conditions $\partial F_M/\partial \phi_j = \partial F_M/\partial \lambda = 0$ deliver a nonlinear system of algebraic equations, which can be solved by Newton (or quasi-Newton) iterations. The perturbative estimates for the condensate wavefunction in the limits $g \to 0$ and $g \to \infty$ are a good starting point for this nonlinear iteration procedure.

Since the same programme can be pursued by using a complete set of basis functions, one may wonder why one chooses finite elements. From a theoretical point of view finite elements offer several nice properties, such as the preservation of positivity and coercivity [231] of the energy functional. In addition, linear elements are known to offer second-order accuracy in variational calculations. From a practical point of view finite elements offer the highest degree of physical flexibility, as they can adapt to complicated geometries. Although such geometrical power remains largely undeployed in the above calculation, the type of procedure prospected by Bao et $al.$ could with relative ease be extended to the case of complex geometries by using suitable families of finite elements such as triangles in 2D and tetrahedra in 3D. For similar procedures in plasma physics see Ref. [232].

2.1.3   Boundary eigenvalue methods

As an example of the use of boundary eigenvalue methods in the BEC context we review the early calculation of the BEC ground state by Edwards and Burnett [233]. These authors consider the 1D GPE with a spherical harmonic potential and solve the boundary eigenvalue problem by means of a Runge-

Kutta space-marching method. In units of $l = \sqrt{\hbar/2m\omega}$ and taking $\Phi(r) = A\phi(r)/r$, the radial GPE eigenvalue problem takes the form

$$\frac{d^2\phi(x)}{dx^2} + \left[\beta - \frac{1}{4}x^2 - N\gamma A^2 \left(\frac{\phi(x)}{x}\right)^2\right]\phi(x) = 0 \qquad (33)$$

where $x = r/l$, $\beta = \mu/\hbar\omega$, and $\gamma = 8\pi a$. The solution depends on the parameters A, γ and N at given β, and these are determined from the boundary conditions at $x = 0$ and $x \to \infty$, plus the normalization condition. The boundary conditions are obtained by inspecting the asymptotic behaviour of Eq. (33).

The boundary condition at $x = 0$ is used as the starting point for a direct Runge-Kutta space-marching method to integrate Eq. (33) up to x_{max}, varying A until the Wronskian of the numerical solution with the asymptotic solution changes sign. This sign change indicates that the boundary condition at x_{max} is indeed matched.

This boundary eigenvalue method provides good results for condensates with radial symmetry. Time-dependent extensions have been developed by Adhikari [234] using matrix methods to advance the solution in time. However, these extensions are still restricted to idealized geometries such as cylinders or spheres, for which analytical work can be developed beforehand to formulate the boundary conditions so as to keep the numerics at a minimum. It can be expected that more systematic approaches, based on matrix eigenvalue solvers, will be needed to deal with more complex geometries.

2.1.4 GPE in imaginary time

An appealing alternative to both numerical eigenvalue and semi-numerical boundary-eigenvalue methods is offered by dynamic evolution in imaginary time. By moving from real to imaginary time through a Wick rotation $\tau = it$, the Schrödinger equation is turned into a reaction-diffusion equation in real time. As a result, the ground-state wavefunction can be computed by advancing the diffusion equation in time until local equilibrium is reached.

Let us consider first the case of a linear Hamiltonian operator $\mathcal{H}(x)$ (in 1D for simplicity), when $\hbar \partial_\tau \psi(x, \tau) = -\mathcal{H}(x)\psi(x, \tau)$. The solution can be represented as a superposition of eigenfunctions,

$$\psi(x, \tau) = \sum_n \psi_n(x)e^{-E_n\tau/\hbar} \qquad (34)$$

where a discrete spectrum of eigenvalues E_n has been assumed. If these are sufficiently well separated the solution decays self-similarly after a sufficiently long transient τ_L at a rate dictated by the smallest eigenvalue, that is

$$\psi(x, \tau) \to \psi_0(x)e^{-E_0\tau/\hbar} . \tag{35}$$

The ground-state energy can be estimated as

$$E_0 \simeq -\hbar \left[\frac{\ln\left(\psi(x_p, \tau_2)/\psi(x_p, \tau_1)\right)}{\tau_2 - \tau_1} \right], \tag{36}$$

where $\tau_2 > \tau_1 \gg \tau_L$ and x_p is a generic "probe" location.

The same procedure can also be applied to the GPE, a few iterations being needed at each time step to deal with nonlinearity. The details of the numerical techniques to advance the solution in time will be discussed in Sec. 2.3, when we shall touch on methods to solve the time-dependent GPE. We proceed immediately below to give an illustration of the method.

Following the evolution of the wavefunction in imaginary time is in fact a way of implementing a minimization of the energy functional. Even more powerful procedures, such as the use of conjugate gradient techniques, are available for this purpose (see *e.g.* [235]).

2.1.5 Ground state in optical lattices

The main application of the static GPE is the evaluation of the ground state of the condensate under various confining potentials. For example, Chiofalo *et al.* [236] have applied the imaginary-time version of the modified Visscher algorithm (see Sec. 2.3.1) to compute the ground state of the BEC in a 1D optical lattice with a superimposed harmonic trap. The optical potential is in the form of Eq. (20), $U(z) = U_0 \sin^2(\pi z/d)$ where d is the lattice spacing. The initial guess for the imaginary-time solution is chosen as a sum of 1D Gaussian profiles centred on the lattice sites and modulated by a Gaussian envelope associated with the harmonic trap. That is,

$$\Phi(z) = A \exp(-m\omega_{ho}z^2/2\hbar) \sum_l \exp[-m\omega_h(z - ld)^2/2\hbar] \tag{37}$$

where ω_{ho} and ω_h are the frequencies of the trap and of the harmonic approximation to the lattice-well potential.

The ground state was computed for increasing values of the number of bosons at a fixed barrier height $U_0 = 1.4E_R$, with $E_R = h^2/(8md^2)$ being the recoil energy. Typical ground-state densities are shown in Fig. 14. The imaginary-time evolution of the energy of the system is illustrated in Fig. 15.

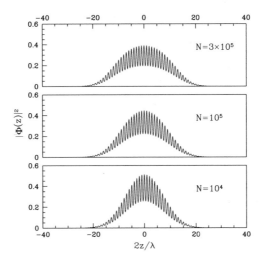

Fig. 14. Condensate in optical lattice plus harmonic trap: density profile for various numbers of atoms. The abscissa is scaled with the lattice period $d = \lambda/2$. Redrawn from Chiofalo *et al.* [236].

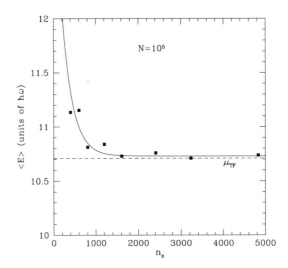

Fig. 15. The average total energy of a condensate with 10^5 atoms in a harmonic trap, as obtained at convergence in the imaginary-time simulation, plotted as a function of the number of grid points. The dashed line shows the value of the Thomas-Fermi chemical potential. From Chiofalo *et al.* [236].

2.1.6 Vortices

A vortex in a superfluid BEC is a topological defect with quantized circulation and vanishing particle density at the core (see Sec. 1.2.4). One of the primary motivations of the Gross-Pitaevskii approach was indeed the study of quantized vortices [220,222].

The equilibrium configuration of a condensate in a spherical trap with a vortex at the centre is obtained by making an Ansatz of the form $\Phi(r)\exp(i\kappa\phi)$ for the vortex wavefunction, where κ are the quanta of circulation. This leads to a modified GPE with a centrifugal term,

$$\mu\Phi(r) = \left[-\frac{\hbar^2}{2m}\left(\Delta_r + \frac{\kappa^2}{r^2}\right) + V_{ext}(r) + g|\Phi(r)|^2 \right]\Phi(r) \tag{38}$$

where Δ_r is the radial component of the Laplacian operator. The numerical solution of Eq. (38) yields a condensate profile with a hole at the trap centre [237], whose size is of the order of the healing length $\xi = \hbar/\sqrt{nmg}$ where n is the average condensate density. In the Thomas-Fermi limit the healing length can be estimated [224] as $\xi/R \sim (a_{ho}/R)^2$ with R the radius of the condensate, and is therefore much smaller than the condensate size.

For typical condensates in a trap the size of the vortex core is far below the experimental spatial resolution. However, in a free expansion of the cloud the vortex core also expands [238] up to a size which becomes detectable by imaging techniques. The vortex state also influences the interference pattern of two expanding condensates: the fringes between a BEC hosting a vortex and a reference BEC should show bifurcations in "Y" shape, contrary to the vortex-free case where the fringes are parallel [239,240].

A vortex created at the centre of a stationary trap is not thermodynamically stable: the vortex state corresponds to a maximum of the energy functional and will tend to spiral out of the trap in a finite time [241]. The instability of the vortex solution is also seen in the excitation spectrum, which shows modes of positive norm and negative frequency corresponding to a precession of the vortex line around the centre of the trap [242,243]. The vortex state can be stabilized by setting the trap into rotation, when it becomes a minimum of the energy functional. One also needs a mechanism to transfer angular momentum to the vortex state, since the condensate is a conserving system, and this is achieved for example by means of a rotating perturbation which is not axisymmetric [244]. The mechanisms of vortex decay are discussed in Sec. 2.3.6.

A value for the critical angular velocity Ω_c of vortex nucleation is estimated from thermodynamic arguments by comparing the energy of a vortex state, which in the rotating frame is $E_v - \Omega L_z$ where Ω is the rotation frequency

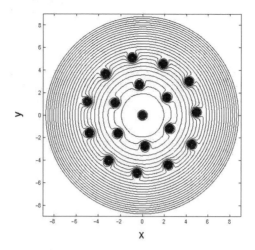

Fig. 16. Vortex lattice obtained by the numerical solution of a 2D GPE for a rotating BEC, from Castin *et al.* [240].

and L_z the angular momentum component along the rotation axis, with the energy E_0 of a condensate without vortices. Since the angular momentum per particle corresponds to $\hbar\kappa$, this argument yields $\Omega_c = (E_v - E_0)/(N\kappa)$, scaling in the Thomas-Fermi limit as $\Omega_c \sim (\hbar/mR_\perp^2)\ln(R_\perp/\xi)$ [224,238]. However, such an estimate fails to reproduce the critical velocity measured in the experiments [116], where vortex nucleation occurs through a dynamic instability (see Sec. 2.3.6).

A rotating condensate is described in the rotating frame by a static GPE including an inertial term,

$$\mu\Phi(\mathbf{r}) = \left[-\frac{\hbar^2}{2m}\nabla^2 + V_{ext}(\mathbf{r}) + g|\Phi(\mathbf{r})|^2 + \Omega L_z\right]\Phi(\mathbf{r}). \qquad (39)$$

The numerical solution of Eq. (39) requires special care in treating the term ΩL_z, as this is not diagonal in position nor in momentum space [240]. Starting with different trial states one may find single-vortex as well as multi-vortex solutions. In the latter case the vortices tend to arrange themselves in a configuration resembling a triangular lattice [244,240] (see Fig. 16).

An important feature which has emerged from the solution of the 3D GPE in a cigar-shaped trap is that vortex lines are bent into a "U" shape [245]. They tend to be parallel to the trap axis in the central region, but exit laterally from the condensate away from the trap centre (see Fig. 17). Such symmetry-breaking states are the minimum-energy solutions of the Gross-Pitaevskii functional for elongated geometries [246], as can also be inferred

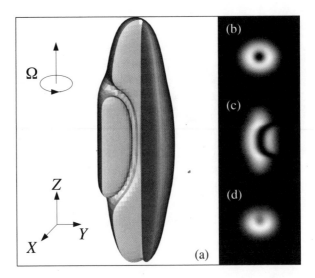

Fig. 17. Vortex bending from the numerical solution of the 3D GPE with the choice of parameters as in the ENS experiments: (a) shows a transverse section and (b) to (d) are the predicted 2D images from various directions (Z, X and Y, respectively). From Garcia-Ripoll *et al.* [245].

from semi-analytic models [247,248]. It has been verified numerically that the "U" shape of the vortex line is preserved in the expansion of the condensate [248].

Vortex bending has been suggested as a possible explanation for the low contrast of the vortex cores in the images obtained in the ENS experiments [245]. A second generation of experiments at ENS has actually observed bending of the vortex lines by simultaneous imaging of the condensate in two perpendicular directions [249]. These experiments have also detected "N" shaped vortex lines which were not met in the numerical studies, being presumably higher-energy configurations.

2.2 Many-body effects

The GPE ignores the presence of any fraction of non-condensed particles around the condensate. In reality, many-body effects can prevent full condensation even at $T = 0$. They start to become important when the dilution parameter is increased and dimensionality is reduced, as in very tight or very flat harmonic traps or in optical lattices at low filling.

It is therefore important to estimate the effects of the noncondensate in the ground state of the gas. To this purpose one needs to solve the N-body

Schrödinger equation. From the computational viewpoint this marks a drastic change as compared to the methodologies discussed so far, since it would require the solution of a partial differential equation in several hundred dimensions. Tackling this task with grid methods is simply unviable: for instance, a cloud of one hundred bosons in 3D, each discretized over only ten grid points, would generate as many as 10^{300} computational degrees of freedom. The way out is provided by grid-free techniques, *i.e.* computational methods in which degrees of freedom are selectively conveyed to strategic portions of phase space where the relevant physics takes place. The method of choice is Quantum Monte Carlo. In particular, from the aforementioned equivalence between the Schrödinger equation and reaction-diffusion systems, Diffusion Monte Carlo is the best option.

2.2.1 Diffusion Monte Carlo

A prominent tool for the study of quantum many-body systems is the Quantum Monte Carlo method (see *e.g.* [250]). The method comes in three main flavours: Variational Monte Carlo (VMC), Diffusion Monte Carlo (DMC), and Path-Integral Monte Carlo (PIMC). The VMC method is the simplest one, but its usefulness heavily hinges on the availability of a good guess for the many-body wavefunction (for an introduction see *e.g.* [229,251]). DMC and PIMC are numerical methods to study the equilibrium properties of many-body systems at zero and at finite temperature, respectively.

Here we shall be concerned mainly with DMC (see *e.g.* [251]), while PIMC will be discussed in Sec. 3.1.4. The primary goal of DMC is the computation of the ground state and of low-lying excited states of the N-body Schrödinger equation. This amounts to solving the linear eigenvalue problem

$$\mathcal{H}(\mathbf{R})\Psi(\mathbf{R}) = E\Psi(\mathbf{R}) \qquad (40)$$

where

$$\mathcal{H}(\mathbf{R}) = -\frac{\hbar^2}{2m}\sum_{i=1}^{N}\nabla_i^2 + V(\mathbf{R}) \qquad (41)$$

is the N-body Hamiltonian in $3N$ dimensional space $\mathbf{R} = (\mathbf{r}_1, \mathbf{r}_2, \ldots, \mathbf{r}_N)$. A central two-body interaction model assuming $V(\mathbf{R}) = \sum_{i<j} v(|\mathbf{r}_i - \mathbf{r}_j|)$ is adequate to account for most relevant physical effects. However, for dilute atomic gases one does not choose for $V(\mathbf{R})$ the available two-body atom-atom potential, because this would lead to a solid-like ground state (see Sec. 1.2.1). The two-body potential is replaced by a pseudopotential which does not support bound states, so that the gaseous BEC becomes the ground state. It has been explicitly verified that the ground state properties of the gas are to a large extent independent of a specific choice for $V(\mathbf{R})$ [252,253].

The DMC method is based on the equivalence between the Schrödinger equation in imaginary time and the diffusion (or Fokker-Planck) equation in real time. Under a Wick rotation the Schrödinger equation turns into a diffusion-reaction equation with diffusivity $D_q = \hbar/2m$,

$$\partial_\tau \Psi(\mathbf{R}, \tau) = D_q \sum_{i=1}^{N} \nabla_i^2 \Psi(\mathbf{R}, \tau) - V(\mathbf{R})\Psi(\mathbf{R}, \tau)/\hbar. \tag{42}$$

The solution is a sum of decaying eigenfunctions $\Psi(\mathbf{R}, \tau) = \sum_n \Psi_n(\mathbf{R}) \exp(-\omega_n \tau)$, and, under some conditions already discussed in Sec. 2.1.4, the asymptotic solution decays self-similarly according to the lowest eigenvalue $E_0 = \hbar\omega_0$,

$$\Psi(\mathbf{R}, \tau) \to \Psi_0(\mathbf{R})e^{-\omega_0 \tau}. \tag{43}$$

The lowest eigenfrequency can then be computed as in Eq. (36). The imaginary-time dynamics thus performs a natural projection onto the lowest-energy eigenstate.

In actual practice a problem arises. Since the norm is not conserved, the signal may be lost in the noise before the lowest eigenvalue is resolved, unless the spectrum is well separated. The remedy is to renormalize the wavefunction at almost every time step, so that the norm remains unitary all along the evolution. Formally this amounts to sustaining the wavefunction with a compensating shift in the potential energy, that is Eq. (42) is changed into

$$\partial_\tau \Psi(\mathbf{R}, \tau) = D_q \sum_{i=1}^{N} \nabla_i^2 \Psi(\mathbf{R}, \tau) + (E_s - V(\mathbf{R})) \Psi(\mathbf{R}, \tau)/\hbar \tag{44}$$

where E_s is tuned so as to keep the norm constant. On reaching the ground state we have $E_s = E_0$, since by definition the energy shift must exactly compensate the ground-state decay.

Initialization is effected by choosing a set of W walkers, which are distributed at random in the $3N$-dimensional space at positions $\mathbf{R} \equiv [\mathbf{r}_1, \ldots, \mathbf{r}_W]$. The diffusion equation is then solved by a classical Monte Carlo sweep in which each move on a time step $d\tau$ consists of:

(a) Diffusion (kinetic energy): move each walker with a Brownian displacement to $\mathbf{r}'_k = \mathbf{r}_k + \mathbf{d}_k$, where $\mathbf{d}_k$ is drawn from a distribution with variance $\sigma^2 = 2D_q d\tau$ (e.g. a Gaussian distribution);

(b) Reaction (potential energy): set $q = \exp((E_s - V)d\tau/\hbar)$ with $V = [V(\mathbf{R}) + V(\mathbf{R}')]/2$ and let the walker survive with probability q if $q < 1$, otherwise $\text{int}(q)$ new walkers are generated;

(c) Normalization: the energy shift E_s is adjusted so as to keep the number of walkers constant. A new cycle is then started until stationary conditions are reached. At the end of the procedure $E_s \to E_0$ yields the ground-state energy, while the wavefunction can be estimated from the positions of the W walkers as $\Psi(\mathbf{R}, \tau) = \langle \sum_k \delta[\mathbf{r}_k(\tau) - \mathbf{r}] \rangle \to \Psi_0(\mathbf{R})$. Here the brackets denote averaging over an ensemble of realizations.

In Coulombic systems the potential is unbounded and the potential energy contribution may lead to a demographic explosion if walkers come too close to each other. This implies large fluctuations in the density of walkers and leads to unacceptably large statistical errors. For Bose condensates a similar divergence does not arise, but it is nonetheless a good practice to resort to importance sampling. In this case the wavefunction is defined with the aid of a (real) guiding function $\Psi_G(\mathbf{R})$, which hopefully is already close to the true ground state. A new "trial" distribution is hence $f(\mathbf{R}, \tau) = \Psi(\mathbf{R}, \tau) \cdot \Psi_G(\mathbf{R})$ Simple algebra shows that Eq. (44) turns into the following Fokker-Planck equation,

$$\partial_\tau f(\mathbf{R}, \tau) = D_q \boldsymbol{\nabla} \cdot [\boldsymbol{\nabla} f(\mathbf{R}, \tau) - \mathbf{U}(\mathbf{R}) f(\mathbf{R}, \tau)] + (E_s - E_l(\mathbf{R})) f(\mathbf{R}, \tau)/\hbar,$$
$$\text{(45)}$$

where $\mathbf{U}(\mathbf{R}) = 2\boldsymbol{\nabla} \ln |\Psi_G(\mathbf{R})|$ is the local drift (sometimes referred to as the "quantum force") and $E_l(\mathbf{R}) = (\mathcal{H}(\mathbf{R})\Psi_G(\mathbf{R}))/\Psi_G(\mathbf{R})$ is the local energy associated to the guiding wavefunction.

The DMC procedure can then be applied to the Fokker-Planck equation (45). The diffusion step becomes a diffusion-advection move, with the inclusion of the drift term $\mathbf{U}(\mathbf{R})$ in the walker displacements. The crucial advantage is that the reaction step is now controlled by the factor $E_s - E_l(\mathbf{R})$. With an optimized choice for the guiding function this energy difference can be minimized and so are the fluctuations in the distribution $f(\mathbf{R}, \tau)$.

2.2.2 Homogeneous Bose gas beyond the dilute limit

The ground-state energy of a dilute gas of hard-sphere bosons was estimated by Lee, Huang and Yang [254] from a second-order perturbative expansion, leading to the result shown in Eq. (8). The next-order correction is logarithmic and of the same order as non-universal terms that depend on the model interaction potential.

The terms beyond mean field start to become relevant as the value of the scattering length can be tuned up in current experiments on dilute gases. In a DMC simulation Giorgini et al. [252] have obtained the ground-state energy of a dilute gas as a function of the diluteness parameter for both a hard-sphere model and a soft-sphere potential well with scattering length comparable to its width. The trial wavefunction is of the Bijl-Jastrow type,

$$\Psi_T(\mathbf{R}) = \prod_{i<j} \psi\left(|\mathbf{r}_i - \mathbf{r}_j|\right) \tag{46}$$

where $\psi(r)$ is the solution of a two-body Schrödinger equation at short distances matched to a constant at large distances. The results of the simulation show that the expression in Eq. (8) gives a reasonable prediction up to $na^3 \simeq 10^{-2}$, where non-universal corrections start to emerge.

A question that remains open concerns the limit where the scattering length a, which is much larger than the size of an atom, is increased to the point where it becomes comparable to the interparticle distance. In this limit the hard-sphere model approaches crystallization into a close-packed structure. A VMC study by Cowell et al. [255] including only two-body correlations shows that the ground-state energy interpolates from the dilute limit given by Eq. (8) to a Fermi-like expression $E/N \propto n^{2/3}$ at large coupling. The latter behaviour is obtained under the assumption that the interparticle distance remains the only available length scale close to a Feshbach resonance when $a \to \infty$.

Of course, the three-body and higher-order terms are non-negligible at strong coupling in an experiment or in a full simulation, yielding an instability towards recombination into molecules. The Efimov states might also play a role [256] close to a Feshbach resonance. The lifetime of a not-so-dilute atomic gas has been estimated by Braaten [257] from perturbation theory. Numerical simulations would require an accurate modelling of the loss channels and seem to be still lacking.

2.2.3 Disordered Bose gas

The study of disordered Bose systems has been the object of intense theoretical and experimental investigations. Among others, major questions relate to the problem of boson localization and to how the nature of the superfluid-insulator transition and the properties of elementary excitations are affected by the presence of disorder (see e.g. [258] and references therein).

Disorder in quantum systems is often modelled by adding a fraction of static impurities, which are randomly distributed and act as a perturbation. An example of disordered Bose fluid is liquid ^{4}He adsorbed in a porous medium [259]. Periodic, quasi-periodic, or disordered arrays of potential wells can be created for an atomic gas by suitable combinations of laser beams and/or speckle [144].

In a dilute gas the particle-impurity coupling is described by a contact pseudopotential involving an s-wave scattering length a_{pi}. Within the Bogoliubov framework it is then possible to estimate analytically the effect of weak disorder on the condensate and superfluid fractions up to second order in perturbation theory [260,261],

$$n_0/n = 1 - \frac{8}{3\sqrt{\pi}}\gamma^{1/2} - \frac{\sqrt{\pi}}{2}\gamma^{1/2}C \qquad (47)$$

and

$$n_s/n = 1 - \frac{2\sqrt{\pi}}{3}\gamma^{1/2}C. \qquad (48)$$

Here we have set $\gamma = na^3$ and $C \equiv c(a_{pi}/a)^2$, c being the impurity concentration. Already at a perturbative level the disorder is more effective in depleting the superfluid than the condensate.

A numerical study of the effects of disorder in BEC fluids at $T = 0$ has been performed by Astrakharchik *et al.* [262]. These authors employ a DMC method to compute the superfluid and condensate fractions in a hard-sphere Bose gas in the presence of a number N_i of static impurities placed at random in a periodic box. Importance sampling is made *via* the use of Jastrow functions and averaging over ten impurity configurations is reported to provide sufficient statistical accuracy for systems of up to 64 bosons. The superfluid fraction is obtained by a zero-temperature extension of the winding-number technique employed in PIMC computations (see Sec. 3.1.2). In practice, it is measured as the ratio $n_s/n = D_s/D_q$ where $D_q = \hbar/2m$ and D_s is the diffusivity of the centre-of-mass. The condensate fraction is obtained from the asymptotic behaviour of the one-body density matrix, see [263].

The main results of this study can be summarized as follows. In the regime of weak disorder ($C \sim 1$) and low density ($na^3 \sim 10^{-5}$) good agreement is found with analytic results from the Bogoliubov theory. In the strong-disorder regime ($C \sim 100$) the numerical simulations yield lower depletions than predicted by the theory and show that the superfluid is depleted more rapidly than the condensate.

Disorder also affects the nature of phase transitions. It changes the phase diagram and the excitation spectrum for a Bose gas in an optical lattice [148]. It also causes a shift of the critical temperature for Bose-Einstein condensation [259,264].

2.2.4   Confined Bose gas beyond the dilute limit

The GPE comes from a local density approximation (LDA) on the 3D energy functional, where only the mean-field term is retained. As the dilution parameter is increased *e.g.* by exploiting a Feshbach resonance, the GPE approximation breaks down. The behaviour of the confined Bose gas in this regime has been investigated by various approaches. A perturbative expansion yields an improved energy functional which contains an LDA term up to second order in the Lee-Huang-Yang expansion *plus* some surface contributions [265]. Since the LDA correction term is positive definite, this approach predicts a greater role for the repulsive interactions than in the simple GPE solution. This has observational consequences *e.g.* on the density profile of the cloud, which should show an increased mean-square radius.

VMC and DMC calculations have been performed in a 3D isotropic trap [253,266–268]. The effect of the harmonic confinement is easily included in the trial wavefunction as a one-body term, so that

$$\Psi_T(\mathbf{R}) = \prod_{i<j} \psi\left(|\mathbf{r}_i - \mathbf{r}_j|\right) \prod_k e^{-\alpha r_k^2} \qquad (49)$$

where $\psi(r)$ is the two-body amplitude as for the homogeneus gas (see Sec. 2.2.2) and α is a variational parameter. For $na^3 > 10^{-3}$ the second-order corrections at the LDA level are found to be in good agreement with the results of the simulation, while neglecting the terms beyond mean-field would overestimate the peak density by 20%. This in turn implies that the use of the Gross-Pitaevskii model in extracting the scattering length from the measured mean-square radius of the cloud overestimates it [269].

VMC and DMC schemes have also been used to estimate the density profile and the condensate wavefunction of a hard-sphere Bose gas in the strongly interacting regime up to $na^3 \sim 0.1$ [266,268]. Full depletion occurs at the centre of the trap and the BEC becomes localized at the boundary, in close analogy with ^{4}He droplets [270]. A hypernetted chain approximation has also been applied to the calculation of density profile, momentum distribution, and static structure factor of strongly interacting hard-sphere and soft-sphere Bose gases [271,272].

(i) One-dimensional confined gases. In reduced dimensionality the effect of increasing the interaction strength is more dramatic and leads to qualitatively new effects. Systems of N bosons interacting with a contact potential in 1D are predicted to undergo fermionization at increasing coupling strength, reaching the Tonks limit [273] where an exact mapping holds between the many-body-ground-state wavefunction $\Psi_B(\mathbf{R})$ of the Bose gas and the ground state $\Psi_F(\mathbf{R})$ of the ideal gas of spinless fermions. That is

$$\Psi_B(\mathbf{R}) = |\Psi_F(\mathbf{R})|, \qquad (50)$$

as demonstrated by Girardeau [274–276]. Boson-fermion mapping applies also to the excited states of the gas and hence to excitation spectra and to thermodynamic properties.

In the Tonks limit the 1D gas is not condensed, as the population of the zero-momentum state is not macroscopic but scales as N^α with $\alpha < 1$. Similarly, while the ground-state energy of a gas of N weakly coupled bosons scales linearly with N, in the Tonks gas the ground-state energy scales like N^2, as for fermions obeying the Pauli exclusion principle. Tonks-Girardeau gases have not been observed as yet, although quasi-1D Bose gases are currently realized in highly elongated traps with an aspect ratio of order 100 [277].

While in the homogeneous case the 1D Bose gas has been exactly solved by Lieb and Liniger [278], in harmonic confinement one needs to resort to approximate schemes [279,280] or to numerical simulations [281,282,262]. A primary question is what aspect ratios and particle densities are needed in practice for genuine 1D behaviour to set in DMC simulations do not rely on any form of coordinate separation or local-density assumptions and are therefore well placed to address the problem of fermionization from a general perspective.

DMC studies of the transition from 3D to 1D have been performed in a cigar-shaped trap by varying its aspect ratio L [281,282]. The calculations become increasingly demanding with growing aspect ratio, since the time step scales with the magnitude of the harmonic oscillator length in the tightly confined direction and hence with the inverse square root of L. The DMC results show that at low L the gas behaves bosonically, but exhibits a clear transition to fermionic behaviour along the axial direction for $L > 1000$ (see Fig. 18). Fermionization is apparent in the contribution to the energy from the reduced 1D problem, where a smooth crossover is found from the 3D Gross-Pitaevskii to the 1D Lieb-Liniger regime [282].

The momentum distribution of the trapped Tonks gas has also been obtained by approximate theoretical and DMC methods [283,284] and its asymptotic-tail behaviour is known analytically [285].

2.2.5 Superfluid to Mott-insulator transition in atomic gases

Considerable attention has been paid to the problem of determining the ground-state phase diagram of correlated bosons on a lattice, as described by "Bose-Hubbard" Hamiltonians [148]. In the limit of high lattice barriers and low tunnelling rates, Bose-Einstein condensates trapped in an optical lattice at low filling provide an especially interesting instance of lattice boson systems.

The Bose-Hubbard model Hamiltonian reads

$$\mathcal{H} = -t \sum_{\langle i,j \rangle} \hat{a}_i^\dagger \hat{a}_j + \frac{1}{2} U \sum_i \hat{n}_i^2 - \sum_i \varepsilon_i \hat{n}_i, \qquad (51)$$

where $\hat{a}_i^\dagger$ and $\hat{a}_i$ are the boson creation and destruction operators on a lattice

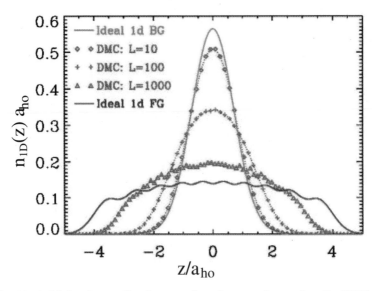

Fig. 18. Axial density profile of a gas of ten bosons: the results of a DMC study (symbols) for increasing aspect ratio L are compared with the asymptotic limits of an ideal Bose gas and an ideal Fermi gas (solid lines). Adapted from Blume [281].

site i and $\hat{n}_i = \hat{a}_i^\dagger \hat{a}_i$ is the site occupation number operator. The parameters are the first-neighbour hopping matrix element t, the on-site repulsion U, and the on-site energy ε_i which is site-dependent in the presence of external confinement. In the Hamiltonian (51) there is a competition between the kinetic-energy tunnel term and the interaction-energy repulsive term. At large tunnelling rates the system is superfluid, while for strong repulsions and low filling the particles are localized at each lattice site and a Mott-insulator phase is found. In the macroscopic limit the insulating phase is characterized by commensurate filling of the lattice sites with zero compressibility and a gap in the excitation spectrum. The superfluid - Mott insulator transition in a Bose gas on a lattice is an example of "quantum phase transitions", which are controlled by changing some parameter in the Hamiltonian rather than some thermodynamic variable (see *e.g.* the book of Sachdev [286] and references therein).

The superfluid-insulator transition has been observed in a pioneering experiment [149] on a gas of ^{87}Rb atoms confined by a combination of a 3D optical lattice and a harmonic trap (see Sec. 1.3.5). The harmonic confining potential has been found to generate a series of new physical effects, having no counterpart in an unconfined system. Theoretical investigations have addressed the problem at mean-field level [150,287–289]. Exact diagonalization has been performed on small systems, where the superfluid fraction has also been estimated [290]. Numerical simulations have determined the density profile and the momentum distribution of the cloud using the Worm algorithm [291] (see Sec. 3.2.1) or the world line Quantum Monte Carlo method [258,292] (see Sec. 3.1.4).

Local observables need to be defined owing to the lack of any traslational invariance from the presence of the confining trap. The main ones are the local density $n_i = \left\langle \hat{a}_i^\dagger \hat{a}_i \right\rangle$ and the local compressibility $k_i = \partial n_i / \partial \mu_i = \beta(\langle \hat{n}_i^2 \rangle - \langle \hat{n}_i \rangle^2)$, where $\mu_i = \partial E / \partial n_i$ is the local chemical potential. A series of plateaus appears on following the density profile across the transition, first at the centre of the cloud and then in rings around it, indicating the presence of incompressible regions alternating with superfluid ones. This result, which can be understood from an LDA on the phase diagram of the macroscopic system [289], shows that the superfluid-Mott insulator transition inside the trap loses its global character. This is also evident from the number fluctuations at each site, which have a completely different behaviour from what is found in the macroscopic system. While in the unconfined case the global compressibility diverges as a commensurate filling is approached, in the confined case the local compressibility tends to vanish. This implies that the Mott insulating phase does not proceed along the traditional path of quantum phase transitions [258]. The size of the incompressible regions in the phase diagram has also been calculated and is found to shrink quite rapidly as filling increases.

2.3 Excitations and transport

As for ordinary states of matter, the transport properties of Bose-Einstein condensates can be assessed by measuring the response to time-dependent external loads. Collective modes can be excited by applying weak periodic fields of suitable spatial symmetry and yield information about the strength of the interactions and the superfluid properties of the gas. The nonlinear response to strong perturbations is even richer: a host of new phenomena, such as the emergence and the evolution of coherent structures (solitons, vortices, skyrmions) characterize the behaviour of BEC fluids far from equilibrium.

The main theoretical tool to explore transport in a dilute gaseous condensate is the time-dependent Gross-Pitaevskii equation (TDGPE), reported in Eq. (26). The solution of the TDGPE has yielded good agreement with experiment for a number of phenomena: spectra of collective excitations under harmonic or optical-lattice confinement, interference patterns, transport and tunnelling through an optical lattice, vortex dynamics, soliton propagation, shock-wave dynamics, four-wave mixing, atom-laser output, and expansion of rotating condensates.

The TDGPE describes the dynamics of a coherent condensate at the mean-field level and omits dissipation. It cannot deal with the dynamics of the cloud when quantum or thermal depletion play an important role. Examples of such situations are condensate formation and decay, phase decoherence, damping of collective excitations, vortex formation, and excitations from a non-mean-field ground state - such as in the Mott insulator, or for a rapidly rotating cloud, or in a low-dimensional Bose gas at strong coupling. The theoretical approaches in these cases span from exact diagonalization for small systems to the solution of the equation for the Wigner distribution function and to time-dependent density functional theory (see *e.g* [293]).

2.3.1 Approaches to solve the time-dependent GPE

A distinction must be made at the start between explicit and implicit methods, as for the numerical solution of any time-dependent differential equation. Explicit methods compute the solution at time $t + dt$ at spatial location $\mathbf{x}$ from only the solution at time t in a local neighbourhood of $\mathbf{x}$, that is

$$\psi(t + dt) = (1 - iH dt/\hbar)\, \psi(t) \tag{52}$$

where $\psi(t)$ is the state vector associated with a discrete representation of the wavefunction $\psi(\mathbf{x}, t)$ and H is the corresponding matrix representation of the Hamiltonian operator. Since this is a differential operator, the matrix representation H in real space is sparse, *i.e.* has only a few non-zero elements around the diagonal. Explicit methods are very efficient on a per-time-step basis, because the product $H \cdot \psi$ involves only a few operations to update the value of the unknown field at each space location. However, many steps are required to achieve the steady state because of the stability restriction on the time step,

$$\omega_M dt < 1 \tag{53}$$

where ω_M is the largest eigenvalue of H [294]. In addition, a naive application of explicit methods to Schrödinger-like equations leads to unconditionally unstable discretizations. In other words, any (however small) time step leads to disruptive instabilities because unitarity is violated. The aforementioned instabilities can be cured by resorting to more sophisticated space-time discretizations, which are designed so as to preserve unitarity [295].

An alternative to explicit methods is to apply implicit Crank-Nicholson methods. Implicit methods evaluate the right-hand side of the evolutionary equation by using a blend of present states (time t) and future states (time $t + dt$). The Crank-Nicholson method is based on the simple trapezoidal rule

$$(1 + iH dt/2\hbar)\, \psi(t + dt) = (1 - iH dt/2\hbar)\, \psi(t)\,. \tag{54}$$

The main appeal of implicit methods is that they relax the stability condition because the norm of the time propagator $T(dt) = (1 + iH dt/2\hbar)^{-1}(1 - iH dt/2\hbar)$

is always smaller than unity. As a result, implicit methods can march to the steady state by large steps. Note that $(1 + iHdt/2\hbar)^{-1}$ is a full matrix even though H is very sparse, so that the problem becomes global as is implied by the violation in causality associated with Eq. (54). This is a computationally intensive task, whose cost has to be weighted against the reduction in the number of time steps. Many GPE solvers are oriented towards finite-difference discretizations combined with implicit Crank-Nicholson time-marching schemes. This leads to numerical matrix algebra techniques, such as Alternate Direction Implicit methods [294], for the solution of the corresponding linearized matrix problem.

(i) Explicit finite-difference methods: Modified-Visscher method. An elegant explicit method for the solution of a linear Schrödinger equation was proposed in the work of Visscher [296]. The starting point is to express the wavefunction as the sum of its real and imaginary parts, $\psi = A + iB$, and then to cast the Schrödinger equation in a symplectic form,

$$\partial_t \begin{pmatrix} A \\ B \end{pmatrix} = \frac{1}{\hbar} \begin{pmatrix} 0 & \mathcal{H} \\ -\mathcal{H} & 0 \end{pmatrix} \begin{pmatrix} A \\ B \end{pmatrix} . \tag{55}$$

This form highlights the analogy with Hamiltonian equations and opens access to the wide body of time marching techniques that are available in the Molecular Dynamics literature [297]. A natural choice is the standard leapfrog scheme,

$$\left. \begin{aligned} A_i^{2n} &= A_i^{2n-2} + 2\, H_{ij}^{2n-1} B_j^{2n-1} dt/\hbar \\ B_i^{2n+1} &= B_i^{2n-1} - 2\, H_{ij}^{2n} A_j^{2n} dt/\hbar \end{aligned} \right\} \tag{56}$$

where the indices i, j refer to the space discretization ($x_i = idx$, say) and the index n to the temporal one ($t_n = ndt$). The specific form of the matrix H_{ij} depends on the details of the space discretization. It is convenient to split it into the kinetic and potential energy components, $H_{ij} = K_{ij} + V_{ij}$. A simple one-dimensional second-order finite-difference scheme would yield $K_{ij} = -D_q(\delta_{i,j+1} - 2\delta_{ij} + \delta_{i,j-1})/2dx^2$ and $V_{ij} = V_i \delta_{ij}$. This shows that the kinetic energy matrix is tridiagonal, while the potential energy matrix is diagonal.

A disturbing feature of the leapfrog scheme is the staggered nature of the discrete time sequence: the real part of the wavefunction is only defined at even time steps, while the imaginary part lives only at odd time steps. This leaves some ambiguity as to the proper definition of the probability density $P = |A|^2 + |B|^2$. Two equivalent options are indeed available, $P^{2n} = A^{2n} A^{2n} + B^{2n-1} B^{2n+1}$ or $P^{2n+1} = A^{2n} A^{2n+2} + B^{2n+1} B^{2n+1}$. As shown by Visscher, either choice secures unitarity in time for a time-independent potential. In essence, the numerical instabilities are tamed because the Hamiltonian formulation preserves space-time symmetry.

Since the Gross-Pitaevskii operator depends on time *via* the dependence of the self-interaction on the wavefunction, unitarity has to be carefully checked. To this purpose it proves expedient to move to a synchronized version of the Visscher scheme. The idea is simply to advance A and B in units of two timesteps, using the intermediate time-centred value of B and A. That is,

$$\left.\begin{array}{l} A_i^{n+1} = A_i^{n-1} + 2H_{ij}^n B_j^n dt/\hbar \\ B_i^{n+1} = B_i^{n-1} - 2H_{ij}^n A_j^n dt/\hbar \end{array}\right\} \tag{57}$$

with an Euler start-up ($A_i^1 = A_i^0 + H_{ij}^0 B_j^0 dt/\hbar$, $B_i^1 = B_i^0 - H_{ij}^0 A_j^0 dt/\hbar$).

The synchronized Visscher scheme is very similar to the second-order difference methods proposed by Kosloff for the time-dependent Schrödinger equation [298]. Its application to the GPE was first developed in Ref. [299]. In this work the authors show that, under ordinary Dirichlet or von Neumann boundary conditions, the synchronized Visscher method is unitary also for a nonlinear potential. Its stability is governed by a Courant-Friedrichs-Lewy condition of the form $dt < dt_{free}/(1 + C_1 dt_{free}/dt_V)$, where $dt_{free} = dx^2/D_q$ is the maximum time step allowed by free-particle motion, dx is the mesh size, and $dt_V = V_{max}/\hbar$ is the maximum time step allowed by the potential. In the above C_1 is a numerical constant of order $1/D$, D being the space dimensionality, and V_{max} is the maximum value of the total (self plus external) potential.

As discussed in the original paper, the stability condition is verified if a "weak-potential" condition $V_{max} < \hbar D_q/(C_1 dx^2)$ is satisfied, and this in turn translates into an upper bound on the number of bosons which can be handled with a given resolution dx, $N < C_2 N_g dx/a$. Here N_g is the number of grid points and C_2 is a constant of order unity. In a dilute gas the maximum number of bosons which can be simulated on a grid with N_g points is thus of order N_g. On the other hand, doubling the space resolution implies a fourfold reduction of the time step. As discussed earlier, this is a limitation motivating the use of implicit solvers.

As mentioned in Sec. 2.1.4, real-time algorithms are readily turned into their imaginary-time counterparts for ground-state calculations. The imaginary-time version of the synchronized Visscher method proceeds along the lines discussed above. The starting system is $\partial_\tau A = -\mathcal{H}_{GP} A/\hbar$ and $\partial_\tau B = -\mathcal{H}_{GP} B/\hbar$, showing that the real and imaginary components are coupled only *via* the nonlinear self-interaction. An application to study the ground state of a condensate in an optical lattice [236] has been presented in Sec. 2.1.5.

(ii) Lattice Boltzmann methods for quantum systems. We turn next to discuss an explicit method in which the maximum time step allowed by stability constraints would in principle scale only linearly with the mesh size. This is indeed possible by moving to a mathematical formulation in which the effects of

the kinetic energy are no longer represented by second-order derivatives. This type of representation is used in classical fluid dynamics to solve the Navier-Stokes equations, involving first-order time derivatives and second-order space derivatives, by means of a suitably discretized form of the kinetic Boltzmann equation known as the lattice Boltzmann equation [300]. Leaving details to the original publications [301–303], here we shall just describe the main physical underpinning, namely the principle of adiabatic enslaving.

Consider for simplicity a 1D lattice with right and left walkers, $r(x,t)$ and $l(x,t)$, obeying a first-order hyperbolic dynamics given by

$$\begin{cases} \partial_t r(x,t) + v\partial_x r(x,t) = (r(x,t) - l(x,t))\,/\tau \\ \partial_t l(x,t) - v\partial_x l(x,t) = (l(x,t) - r(x,t))\,/\tau \end{cases} \tag{58}$$

with $\tau > 0$. The left-hand side of Eq. (58) is the standard free-streaming term, whereas the right-hand side represents the linear interaction between the two types of walker. By adding and subtracting the two equations we get

$$\begin{cases} \partial_t S(x,t) + v\partial_x J(x,t) = 0 \\ \partial_t J(x,t) + v\partial_x S(x,t) = -2J(x,t)/\tau \end{cases} \tag{59}$$

where the symmetric component $S = r+l$ is the density and the antisymmetric component $J = r-l$ the current density for the fluid formed by the two species of walker. Since S is conserved while J decays at a rate $2/\tau$ we can solve the second equation under the adiabatic assumption $|\partial_t J| \ll 2|J|/\tau$, which yields $J \sim -v\tau\partial_x S/2$. With this adiabatic result we find

$$\partial_t S(x,t) = \frac{1}{2}v^2\tau\partial_x^2 S(x,t)\,. \tag{60}$$

This is a 1D diffusion equation for the scalar $S(x,t)$ with a diffusion coefficient $D = v^2\tau/2$.

Returning to Eq. (58), it is thus possible to solve a diffusion equation without ever dealing with second-order space derivatives. Of course this comes at the cost of doubling the degrees of freedom, but the gain is often worth the cost since the stability condition for the hyperbolic system (60) is now $dt < dx/v$ instead of $dt < dx^2/D$ as for the diffusion equation.

These ideas can be exported to the context of quantum equations for diffusion in imaginary time, including the case of the GPE, using a lattice Boltzmann equation with two discrete speeds $\pm c$. Multi-dimensional extensions are in principle possible, although they remain to be tested numerically.

(iii) Implicit finite-difference Crank-Nicholson methods. Finite-difference methods have been developed by Adhikari [304] for a radially symmetric conden-

sate, in combination with a Crank-Nicholson time-marching scheme (FDCN). The matrices are tridiagonal in this case since there is only one space variable and consequently they can be solved exactly and efficiently by a purely algebraic version of Gaussian elimination (the Thomas algorithm [294]). Numerical matrix algebra is generally required for multi-dimensional problems.

The burden of matrix algebra can often be considerably reduced by resorting to operator splitting in real space. The basic observation is that at least in simple geometries the Laplace kinetic-energy operator is the sum of three independent second-order derivatives along the three directions of motion. Consider for simplicity the case of Cartesian coordinates in 2D, where $\mathcal{H} = -\hbar^2 \left(\partial_{xx} + \partial_{yy}\right)/2m + V(x,y)$. The matrix problem requires finding the result of the action of the inverse operator $(1 - i\mathcal{H}dt/\hbar)^{-1}$ on a source term. The operator equation $(1 - i\mathcal{H}dt/\hbar)^{-1} = (1 - i\mathcal{H}_y dt/\hbar)^{-1}(1 - i\mathcal{H}_x dt/\hbar)^{-1} + O(dt^2)$ indicates that a two-dimensional matrix problem can be split into the sequence of two one-dimensional matrix problems, to second-order accuracy in the time step dt. Here we define the splitting of the Hamiltonian through the modified splitting method $\mathcal{H}_x = -(\hbar^2/2m)\partial_{xx} + V(x,y)/2$ and similarly for $\mathcal{H}_y$ [305]. In practice this amounts to applying an Alternating Direction Implicit method [294],

$$\begin{cases} \left(1 + i\mathcal{H}_x dt/2\hbar\right)\psi(x,y;t+dt/2) = \left(1 - i\mathcal{H}_y dt/2\hbar\right)\psi(x,y;t) \\ \left(1 + i\mathcal{H}_y dt/2\hbar\right)\psi(x,y;t+dt) = \left(1 - i\mathcal{H}_x dt/2\hbar\right)\psi(x,y;t+dt/2). \end{cases} \qquad (61)$$

Each of these two 1D systems involves simple tridiagonal matrices, which can be efficiently inverted by purely algebraic means.

As for explicit schemes, care is needed in using these methods to handle the nonlinear Gross-Pitaevskii operator in order to prevent the nonlinearity from spoiling unitarity. A practical way to proceed is by a predictor-corrector method. Each time step in Eq. (61) is taken twice: the first step produces a predicted value $\tilde{\psi}$ by evaluating the nonlinear term from $\psi(t)$, and the second uses the updated value $[\psi(t) + \tilde{\psi}(t + dt)]/2$. This preserves second-order accuracy in dt. The scheme also applies to cylindrical coordinates and has been applied to compute the effects of self-interactions on macroscopic tunnelling in optical lattices [306].

(iv) Explicit pseudo-spectral Runge-Kutta method. As an alternative Muruganandam and Adhikari [307] have developed a pseudo-spectral method combined with a four-step explicit Runge-Kutta time-marching scheme (PSRK). The method is based on the expansion of the wavefunction onto a set of interpolating orthogonal functions $h_i(x)$. In 1D for simplicity (but 3D extensions

are conceptually straightforward) one has

$$\Phi(x,t) \sim w(x) \sum_{i=0}^{N-1} \phi_i(t) h_i(x)/w(x_i) \qquad (62)$$

where x_i is a set of collocation points, $\phi_i(t) = \Phi(x_i, t)$ are the unknown amplitudes, and $w(x) = e^{-x^2/2}$ is a weight function. Upon substituting Eq. (62) in the GPE, the second-order derivative receives a matrix representation in which it acts on smooth functions. The GPE is then turned into a set of ordinary differential equations for the unknown $\phi_i(t)$, which are advanced in time with an adaptive fourth-order Runge-Kutta method.

Hermite polynomials are a natural choice for this problem, as they are the eigenfunctions of the linearized GPE in harmonic traps and consequently satisfy the proper boundary conditions by construction. The PSRK has been successfully applied to spherically symmetric, axially symmetric, and anisotropic models, with reported errors of less than 1% for about twenty basis functions. The authors discuss a complementarity of different methods, the PSRK being more suitable for smooth potentials where relatively few and unevenly distributed points are adequate. The FDCN conbined with a large number of uniformly distributed grid points is instead recommended for rapidly varying potentials such as in an optical lattice.

(v) Time-splitting spectral methods. The GPE operator is the sum of purely space-dependent and purely momentum-dependent components. The former is diagonal in a real-space representation and so is the latter in momentum space. One may then think of switching back and forth between these two representations in order to deal only with diagonal matrices. This is indeed possible by using operator splitting techniques in phase space [308].

Splitting of phase-space operators works as follows. Let us take a regular 1D lattice $x_i = i\,dx$ $(i = 1, \ldots, N)$ and its reciprocal wavenumber lattice $k_l = 2\pi l/dx$ $(l = 1, \ldots, N)$. In the real-space representation we first advance the solution using only the potential energy operator to construct $\psi_i = (1 - iV_i dt/\hbar)\psi_i(t)$. Next we transform to momentum space by calculating $\tilde{\psi}_i = \sum_i \exp(-ik_l x_i)\psi_i$ and advance the transformed wavefunction under the effect of the kinetic energy operator to get $\tilde{\psi}_l(t + dt) = (1 + iK_l dt/\hbar)\tilde{\psi}_l$ where $K_l = \hbar^2 k_l^2/2m$. Finally, an inverse transform back to real space delivers the solution at time $t + dt$,

$$\psi_i(t + dt) = \sum_l \exp(ik_l x_i)\tilde{\psi}_l. \qquad (63)$$

Clearly, working with diagonal matrices makes the scheme highly efficient provided the swapping back and forth from coordinates to momenta can be dealt with economically. This is achieved with the Fast Fourier Transform technique, which reduces the $O(N^2)$ complexity of a Fourier transform to a more manageable $O(N \log N)$ complexity.

Other nice features of the time-splitting spectral method are (a) that the spatial accuracy of the spectral representation is virtually exponential, since the Laplace operator becomes k^2 to exponential accuracy; and (b) that the norm of the wavefunction is conserved in time, thus relaxing the stability constraint. For instance, for a condensate in a 1D harmonic trap the time-marching scheme reads $\psi_i = \exp(-iV_i dt/\hbar)\psi_i(t)$ and $\tilde{\psi}_l(t+dt) = \exp(iK_l dt/\hbar)\tilde{\psi}_l$, where $V_i = m\omega^2 x^2/2 + g|\psi_i|^2$. The result is very accurate in space and has second-order accuracy in time, the error in time stemming from operator splitting as a consequence of the fact that the kinetic and potential energy operators do not commute. However, the method is limited to situations where periodic boundary conditions can be applied, for otherwise a spectral representation is no longer viable.

2.3.2 Excitation spectrum of a condensate in harmonic trap

Collective modes of the condensate and more generally of the whole Bose fluid are excited in the laboratory by driving the gas with a time-dependent perturbing field of suitable symmetry. Here we focus on the theory of the excitations of a pure condensate. Calculations on a Bose gas at finite temperature are referred to in Sec. 3.3 and the study of the collective modes in boson-fermion mixtures is reviewed in Sec. 4.3.

The modes of a fluid in a spherical trap can be labelled by the angular-momentum quantum number l and by the quantum number m of the component of the angular momentum along a z-axis. Modes of different l are uncoupled. A third quantum number n labels modes with different numbers of radial nodes. In a cylindrical trap the modes are usually labelled by the same numbers (n, l, m), but the total angular momentum is no longer a good quantum number and the numerical approach requires the solution of sets of coupled equations for different l.

The eigenfrequencies and the density fluctuations of the collective modes of a condensate are calculated from the time-dependent GPE in Eq. (26) by adding a time-dependent drive to the trapping potential. In the linear regime this approach is equivalent to solving the Bogoliubov-de Gennes equations [2], reducing in the homogeneous limit to the Bogoliubov sound waves. The results for the mode frequencies are usefully illustrated by plotting them against the coupling-strength parameter $\eta = Na/a_{ho}$ introduced in Sec. 1.3.2. Figure 19 shows the low-lying quadrupole modes for a ^{87}Rb condensate in the trap geometry of the JILA experiment [72]. The measured frequencies correspond to intermediate values of the coupling strength and are well accounted for by the

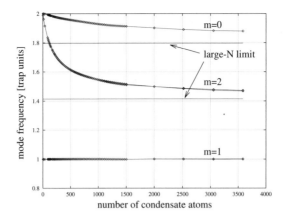

Fig. 19. Low-lying collective modes of a condensate in a cigar-shaped trap plotted against the condensate atom number. From Edwards *et al.* [309].

numerical predictions [309]. A time-dependent variational Ansatz has proved to work well for the monopole and quadrupole modes over the whole range of coupling strength [310].

At small N the quadrupolar mode frequencies in Fig. 19 reduce to twice the bare trap frequency, as is expected for an ideal gas under harmonic confinement. For larger N each frequency tends to saturate to a lower value which depends on the quantum number m and has been estimated analytically as follows [311]. Upon setting $\Phi(\mathbf{r}, t) = \sqrt{n(\mathbf{r}, t)} \exp(i\phi(\mathbf{r}, t))$, the TDGPE can be recast into the combination of a continuity equation and of a quantum Euler equation for the superfluid velocity $\mathbf{v}_s(\mathbf{r}, t) = \hbar \boldsymbol{\nabla} \phi(\mathbf{r}, t)/m$. That is,

$$\partial_t n(\mathbf{r}, t) = -\boldsymbol{\nabla} \cdot [n(\mathbf{r}, t)\mathbf{v}_s(\mathbf{r}, t)] \tag{64}$$

and

$$m\partial_t \mathbf{v}_s(\mathbf{r}, t) = -\boldsymbol{\nabla} \left[\frac{\hbar^2}{2m\sqrt{n(\mathbf{r}, t)}} \nabla^2 n(\mathbf{r}, t) + V(\mathbf{r}, t) + gn(\mathbf{r}, t) + \frac{1}{2} mv_s^2(\mathbf{r}, t) \right]. \tag{65}$$

Equation (65) is the collisionless equivalent of the Landau equation for a pure superfluid. The first term in the brackets is the surface kinetic energy and can be neglected at large N, thus allowing an analytic solution of Eqs. (64) and (65) after linearization.

2.3.3   Interference between condensates

We have briefly reported in Sec. 1.3.3 the observations of interference between expanding condensates by Ketterle *et al.* [101] and the related numerical study of Röhrl *et al.* [103]. These interference effects are regarded as most compelling evidence for a macroscopic wavefunction associated to the broken gauge symmetry in Bose-Einstein condensation. Here we report on an application of the GPE to study interference effects between condensates released from a double well [312]. We also report on the use of the modified Visscher scheme to study the coherent tunnelling of a condensate through a quasi-1D optical lattice, with reference to the experiment of Anderson and Kasevich [89] presented in Sec. 1.3.5.

(i) Expanding condensates. In the work of Wallis *et al.* [312] the imaginary-time GPE is used to generate a realistic initial condition for the numerical integration of the real-time GPE. The latter employs an approximate factorization of the wavefunction along the different directions, so that one solves with significant computational savings a sequence of three 1D GPE's coupled through the self-consistent potential. Of course, the product Ansatz does not allow for the dimensional tangling effects which are inevitably generated in the course of the evolution. However, the authors maintain that these do not affect the ratio of the expansion rates along the different directions.

Interference effects were explored by inspecting the Wigner function associated with the BEC wavefunction provided by the numerical solution of the GPE, that is

$$f_p(x,t) = \frac{1}{2\pi} \int dr \Phi^*(x - r/2, t)\Phi(x + r/2, t) \exp(-ipr/\hbar) . \qquad (66)$$

The time evolution of the fringe patterns in the Wigner function for an interacting condensate is shown in Fig. 20.

(ii) Drop emission from an optical lattice. The modified Visscher scheme was first used to study transport of a condensate through a periodic quasi-1D potential under the drive of a constant force F [313]. A coherent drop of matter is emitted every time that the condensate reaches the edge of the Brillouin zone, if the lattice barrier height is chosen so as to permit tunnelling through the lattice and into vacuum. The drops are separated in time by the period $T_B = h/(Fd)$ of Bloch oscillations, with d the lattice spacing (see Sec. 1.3.5). The results for a non-interacting condensate are shown in Fig. 21. In the numerical study the period of drop emission is independent of the barrier height and of the interactions, as predicted by band-structure theory, but these determine the shape and size of the drops. The relationship between this type of coherent transport and the tunnel of Cooper pairs through a single Josephson junction is discussed in Sec. 2.3.5.

The transport behaviour of the system as a function of its many governing parameters can be summarized in a simple two–dimensional diagram, which reports the fractional number N_{drop}/N of atoms in the first drop as a function

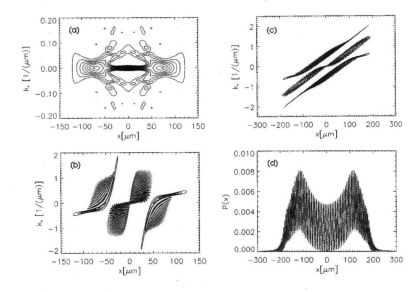

Fig. 20. Wigner function in phase space for 5×10^6 interacting bosons expanding after release from a double well. (a) Initial state; (b) early stage of the expansion at $t = 1.3\,\text{ms}$ (note the changed momentum scale); (c) at $t = 538.2\,\text{ms}$, the fringes near $x = 0$ are displayed with a much higher amplitude due to the earlier overlap of the densities; (d) the corresponding spatially resolved fringes are displayed with a very narrow spacing in the position distribution. From Wallis $et\,al.$ [312].

of the ratio g/g_s, where $g = F/m$ and g_s is the value of g such that the constant drive plus the self-interactions exactly balance the lattice barrier. This diagram is reported in Fig. 22 and offers a useful synthetic view of the phenomenon, even though it is not fully universal. In particular it shows the existence of a threshold for drop emission, which is about $g/g_s \sim 0.15$ for $g = 981\ cm/s^2$ and about half that when g is doubled. A continuous current appears well below the classical threshold at $g/g_s = 1$.

2.3.4   Band structure in optical lattices

The modified Visscher scheme has also been used to probe the energy bands of a condensate in a quasi-1D lattice [133]. A variety of methods have been simulated, that could be realized in the laboratory to probe the band excitation energies and the momentum distribution. These include "kicking" and "shaking" perturbations based on different drives.

Kicking consists of imparting an impulsive velocity $v = \hbar k/m$ to the BEC at time $t = 0$ by imposing a phase shift kz on the ground-state wavefunction: the dynamics of the wavefunction is then followed in time and its Fourier spectrum

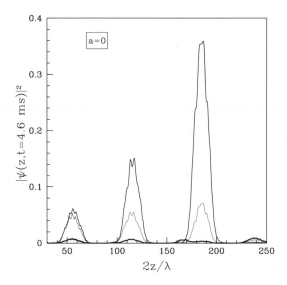

Fig. 21. Density profile of the drops emitted from a non-interacting condensate after 4.6 ms as a function of scaled distance $2z/\lambda$ and for various heights of the lattice barrier U_0 (from largest to smallest, $U_0/E_R = 0.7$, 1.4 and 2.1). The main condensate lies around $z = 0$ and has been subtracted. From Chiofalo et al. [313].

is analyzed so as to detect resonances as a function of the reduced wavenumber q in the Brillouin zone. In the shaking scenario a parametric drive is applied, given by a potential $U_p(z,t) = AU(z)\cos(kz - \omega t)$ over a finite time interval. A static scenario is also explored, in which a phase modulation is imposed on the ground state so as to generate an overlap with a Bloch state of quasi-momentum $\hbar q$: the later energy decay shows a plateau at $E_n(q)$ before return to the ground state. These three methods are found to provide the same energy bands within numerical accuracy and the results are shown in Fig. 23 for the three lowest bands.

The inset in Fig. 23 shows an enlarged view of the lowest band up to $qd/\pi = 0.5$, where the band starts to bend over. On this scale the estimated error of the simulation becomes visible. As already remarked in Sec. 1.3.5, the use of the GPE leads to a quadratic dispersion near the centre of this band.

2.3.5 Josephson-type oscillations and decoherence in a lattice

Let us consider next the current $\bar{p}(t)$ which is carried in a linear semi-classical regime by a periodic condensate as it is driven by a time-dependent external force $F(t)$ through the states of the lowest energy band of a quasi-1D lattice. The current is obtained from the expectation value of the momentum oper-

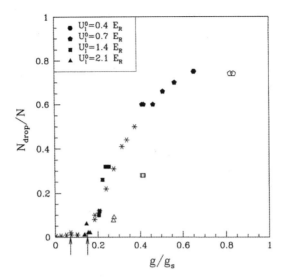

Fig. 22. Diagram for drop emission: the fractional number of particles in the first drop is plotted against scaled gravity g/g_s for $g = 981\,cm/s^2$ (filled symbols) and $g = 1962\,cm/s^2$ (empty symbols), for a variety of choices of system parameters. From Chiofalo et al. [313].

ator $-i\hbar\partial_x$ applied to the Bloch state $Z_{0q(t)}^{+}(x)$, where the time-dependent wavenumber $q(t)$ is related to the external force by $\hbar\dot{q}(t) = F(t)$ and the index $n = 0$ denotes the lowest energy band [131,132]. In the case of a constant force one recovers the Bloch oscillations of the condensate that were observed in the experiments of Anderson and Kasevich [89] and discussed in Sec. 2.3.3. Here we discuss the more general case when the driving force is the sum of a constant term F and a harmonic term $-m\omega^2 x$, yielding

$$\hbar q(t) = Ft + m\omega A \sin(\omega t + \phi_0) \tag{67}$$

where ϕ_0 and A are to be determined from the initial conditions. The case of a purely harmonic external force describes the setup in the experiments of Burger et al. [126] that we described in Sec. 1.3.4.

From the Wannier representation of the Bloch states given in Eq. (21) one easily finds

$$\bar{p}(t) = \frac{\hbar}{d} \frac{\sum_{l\geq 1}[-2\partial g_0(ld)/\partial l]\sin[lq(t)d]}{1 + 2\sum_{l\geq 1} g_0(ld)\cos[lq(t)d]} \tag{68}$$

where d is as usual the lattice spacing, l is a positive integer from labelling of the lattice sites, $g_0(ld) \equiv \int dk\,|f(k)|^2 \cos(kld)$, and $f(k)$ is the Fourier transform of the Wannier function $w_0^{+}(x)$ [145]. From Eqs. (67) and (68) it is immediately evident that the average current driven by a constant force in the

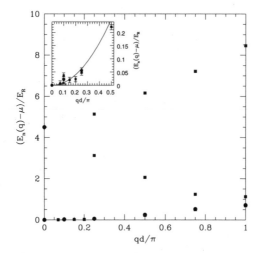

Fig. 23. The three lowest energy bands in the Brillouin zone as obtained by different drives of the condensate (squares, kicking; circles, shaking; triangles, static phase modulation). The inset zooms on the lowest band, where the size of the error bars reflects a two-digit accuracy in energy differences. The solid line shows a quadratic dispersion with the bare atom mass. From Chiofalo et al. [133].

regime that we are discussing is a periodic function of time, with a structure which arises from the relative phases between the lattice wells. The details of the periodic potential and of the interparticle coupling enter through the weighting function $g_0(ld)$.

In the more general case one may insert Eq. (67) into Eq. (68) and perform a multi-mode expansion of both the numerator and the denominator through the use of the Bessel functions $J_n(lAd/a_{ho}^2)$ where a_{ho} is related to the frequency of the harmonic drive by $a_{ho} = \sqrt{\hbar/m\omega}$. The result displays the possibility of multi-mode oscillations with resonances occurring whenever the constant force is tuned to match the resonance condition $\omega_B = n\omega$, with $\omega_B = |F|d/\hbar$ being the frequency of Bloch oscillations. At each resonance the pattern of the current suddenly changes, displaying a number n of sub-oscillation peaks within each fundamental Bloch period. It turns out that there is a range of realistic system parameters for which the current reduces to

$$\bar{p}(t) = p_0 \sum_n (-1)^n J_n(lAd/a_{ho}^2) \sin[(\omega_B - n\omega)t - n\phi_0]. \qquad (69)$$

Equation (69) is formally equivalent to the expression for the current passing through a Josephson weak-link junction between two superconductors when a constant voltage drop $\Delta\mu = Fd$ and an oscillating voltage of strength $U = m\omega^2 Ad$ are applied to it. We can therefore expect that coherent phenomena

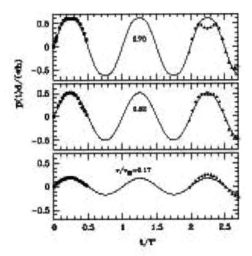

Fig. 24. Multi-mode oscillations of a BEC driven by a harmonic force through a lattice with barrier height $U_0 = 1.5E_R$: average current as a function of scaled time t/T^* with $T^* = 1.1\,T_B$ from an effective-mass correction, for various values of the maximum velocity v attained by the center-of-mass (in units of $v_{BZ} = \pi\hbar/(md)$). From Chiofalo and Tosi [145].

typical of Josephson junctions between superconductors should be observable when a superfluid condensate is driven through an optical lattice in the linear-transport regime.

Figure 24 compares the predictions of Eq. (68) for the oscillations of an infinitely extended BEC driven through a lattice by a harmonic force (solid curves) with the results of a numerical solution of the time-dependent GPE (dots) for a condensate of limited extent which is confined in a harmonic well superposed onto a periodic potential (see Fig. 14 for relevant illustrations of the equilibrium density profiles). The strength of the force determining the parameter A in the theory is characterized by the maximum velocity v reached by the centre-of-mass of the condensate in its oscillations. There is no trace of decoherence in the range of v/v_{BZ} illustrated in the figure, with $v_{BZ} = \pi\hbar/(md)$ being the velocity corresponding to the quasi-momentum at the Brillouin zone edge. Multi-mode behaviour is emerging at $v/v_{BZ} \simeq 0.7$ and tends to be overestimated by the simplified Eq. (69) (triangles). Figure 25 illustrates the behaviour predicted by the numerical solution of the GPE for the condensate wavefunction and for the momentum distribution in this transport regime.

Decoherence sets in at higher centre-of-mass velocities, prior to the onset of Bragg scattering at $v/v_{BZ} = 1$. This is seen from the GPE as a fragmenta-

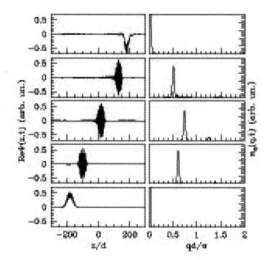

Fig. 25. Real part of the BEC wavefunction (left panels) and of the momentum distribution (right panels) for the condensate in the top panel of Fig. 24 at various times ($t/T^* = 0$, 0.15, 0.27, 0.39, and 0.50 from bottom to top). From Chiofalo and Tosi [145].

tion of the condensate into mutually incoherent parts (see Fig. 26). Above a threshold velocity the condensate breaks up into an incoherent assembly of subsystems randomly moving in the various lattice wells, and at the same time the momentum distribution crumbles away and spreads over the whole Brillouin zone [315]. This behaviour is reminiscent of dynamical localization and quantum chaotic behaviour [316]: the Hamiltonian that governs the system can be traced back to that of a periodically kicked rotator, in which a simple oscillator comes into resonance with an external drive and dynamical chaos sets in after the activation of many degrees of freedom.

As discussed in the work of Burger et al. [126], the predictions from the GPE in the linear-transport regime are in full quantitative agreement with the experimental data on the behaviour of a condensate driven to oscillate through an optical lattice by a harmonic force. However, at stronger drives there are clear differences between the observed behaviour of the real condensate and the emergence of decoherence as predicted by the numerical solution of the time-dependent GPE. As already discussed in Sec. 1.3.4, breakdown of superfluidity in the real condensate seems to occur through the onset of dissipative processes. These are, of course, absent in a mean-field theory.

(i) Extension to higher dimensionalities. Polini et al. [317] have shown that under suitable conditions the Hamiltonian of a gas of cold bosonic atoms inside a rotating lattice in dimensionality $D \geq 2$ can be mapped into that of a

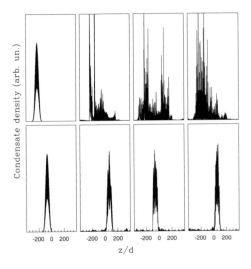

Fig. 26. Superfluid regime (bottom) and decoherence regime (top) for a GPE condensate driven by a harmonic force to oscillate through a periodic potential. The pictures are taken every 50 ms: the initial displacement of the centre of mass is by 27 μm in the bottom strip and 89 μm in the top strip. From Chiofalo *et al.* [314].

frustrated Bose-Hubbard model, the role of magnetic frustration being taken by the angular frequency of rotation of the lattice. In particular, in the limit of large on-site occupation the rotating BEC can be mapped into an array of Josephson junctions in a magnetic field.

Numerical calculations show that the system can be fully frustrated at experimentally attainable values of the frequency of rotation, when a superlattice of vortices is predicted to appear in the ground state. This has several implications on observable quantities such as the superfluid density and momentum distribution. Cold bosonic atoms in a rotating optical lattice promise therefore to provide an ideal condensed-matter system in which to study the effects of frustration in the absence of disorder.

2.3.6 Dynamics of vortices and vortex decay

The dynamics of vortex lines in quantum fluids dates back to studies on superfluid ^{4}He. In atomic Bose-Einstein condensates several aspects of vortex dynamics have been so far investigated [121]: these cover the possible mechanisms for vortex nucleation, the small oscillations of vortex lines and vortex lattices, and vortex decay.

(i) Vortex nucleation and vortex lattice formation. As mentioned in Sec. 2.1.6, the critical angular frequency for vortex generation by stirring a condensate in the ENS experiments [116] is substantially larger than the thermodynamic estimate. It is now understood that in this type of experiment vortices are nucleated through a dynamical instability. For given modulation amplitude and stirring frequency the condensate first distorts itself into a rotating ellipse which is stationary in the rotating frame [318], and then an intrinsic instability transforms this state into a more axisymmetric one hosting some vortices [319]. Surface modes are resonantly excited by the drive near the instability, leading to surface ripples which develop into vortex cores. An estimate of the critical rotation frequency can be obtained from a Landau criterion (see Sec. 1.3.4), *e.g.* by looking at when surface excitations of frequency Ω_l and quantum number l become energetically favourable in the rotating frame [320,248]. This yields

$$\Omega_c = \min_{\{l\}} \left(\Omega_l / l \right). \tag{70}$$

A complete picture of the process of vortex nucleation and vortex-lattice formation has been obtained from numerical simulations of two different kinds. The first uses a time-dependent GPE with a phenomenological dissipation term [321],

$$\hbar(i - \gamma)\partial_t \Phi(\mathbf{r}, t) = \left[-\frac{\hbar^2}{2m}\nabla^2 + V_{ext}(\mathbf{r}) + g|\Phi(\mathbf{r}, t)|^2 - \Omega L_z \right] \Phi(\mathbf{r}, t). \tag{71}$$

Inclusion of dissipation is essential to recover a stable vortex-lattice configuration, since the lattice is a local minimum of the total energy and dissipation is needed to remove the energy released when the lattice is formed. A microscopic modelling of the dissipation term γ has been given in [322]. The numerical solution of Eq. (71) yields a value for the critical angular frequency of vortex nucleation which is very close to the measured one. It also predicts vortex-lattice formation at sufficiently large stirring frequencies.

A second kind of numerical simulation uses a classical field formalism (see Sec. 3.4.2). The time-dependent GPE is viewed as an equation of motion for the whole classical field, which can be decomposed into the sum of the condensate wavefunction and a "noise" term of quantum or thermal origin. In practice, one introduces the required dissipation mechanism by adding at the start of the simulation a randomly distributed noise. Vortex nucleation and vortex-lattice formation are again in good agreement with experiment [323]. An interesting outcome is the prediction of two distinct mechanisms of vortex-lattice nucleation: at "zero" temperature several vortices can enter the cloud at the same time, while at finite temperature the vortices enter one at a time (see Fig. 27).

(ii) Vortex formation in flows past obstacle. Formation of vortices has been

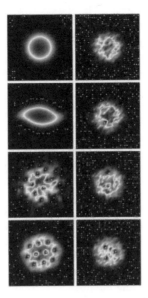

Fig. 27. Vortex lattice formation at $T = 0$ (left panels) and at $k_B T = 8\hbar\omega$ (right panels), as obtained by the classical field method. The panels from top to bottom show subsequent stages in the time evolution of the cloud under a rotating drive (initial state, near the instability, turbulent regime, and stationary vortex lattice). From Lobo *et al.* [323].

inferred in an experiment where a laser beam is moved through a condensate [124] (see Sec. 1.3.4). The idea of studying the origin of drag and dissipation in quantum fluids and the ensuing loss of superfluidity dates back to Feynman's suggestion that the onset of dissipation may be a consequence of vortex shedding from solid obstacles. In an atomic gas one can study dissipation phenomena in an almost pure condensate. Jackson *et al.* [324] have solved the GPE in the presence of an external potential which mimics the presence of a moving obstacle. These authors consider a 2D Gaussian barrier that moves with velocity U along the y axis. They first calculate the wavefunction with a static "object" at the origin and then generate a disturbance which moves through the condensate. This eventually leads to nonlinear topological excitations when U exceeds a critical speed, which lies below the speed of sound in analogy with experimental observations. The process of vortex formation as a result of the accumulation of phase slips is displayed in Fig. 28.

(iii) Dynamics of vortex lines. A vortex state can be viewed as a macroscopically excited Bose-Einstein condensate and its normal modes can be put into resonance with perturbing drives. A first type of excitation for a single vortex line are the Kelvin modes, corresponding to the bending of the line like a guitar string. Since the superfluid velocity field winds around the vortex core in a parity-violating pattern, these modes are chiral, *i.e.* they always rotate in the

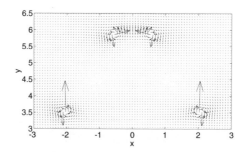

Fig. 28. Plot of the fluid velocity $\boldsymbol{v_s}$ in the vicinity of an object. The arrow length is $40v_s$ in harmonic oscillator units. From Jackson *et al.* [324].

sense opposite to the vortex velocity field. Indications of the creation of Kelvin modes in experiments have been obtained by observing different lifetimes for co-rotating and counter-rotating quadrupole modes in a condensate with a vortex [325]. The shorter lifetime of the counter-rotating modes is due to their Beliaev decay into pairs of Kelvin modes [326], which is instead forbidden by symmetry for the co-rotating modes.

Another family of low-energy modes in a BEC are the shear modes of a vortex lattice, known as the Tkachenko modes [326,327]. These have been observed by applying a transverse perturbation to the vortex lattice [328]. Their frequencies have been estimated by means of the hydrodynamic theory of rotating superfluids with inclusion of the elasticity of the vortex lattice, finding good agreement with experiment [329,330].

(iv) Dynamics of a giant vortex. Vortices with a winding number larger than unity are energetically disfavoured and are not expected to persist if created in a rotating BEC. A number of methods have nevertheless been examined to overcome this instability against vortex dissociation. In fact, giant vortex structures have been created in a rapidly rotating BEC under harmonic confinement by using a near resonant laser beam to produce a density hole surrounded by a large vorticity [331].

Simula *et al.* [332] have carried out numerical simulations of giant vortex structures in a rapidly rotating BEC within the GPE approach. They reproduce the main qualitative features of the observed giant-vortex dynamics, *i.e.* the oscillations of its core area, the formation of a toroidal density hole, and the precessing motion of the vortex line. They are also able to simulate the transverse Tkachenko modes of the vortex lattice.

(v) Vortex decay. Among topological defects of Bose-Einstein condensates the vortices are relatively long-lived. Whereas solitons can decay at any point of the fluid [333], a vortex decays only when it reaches the boundaries since its topological charge is a conserved quantity.

A vortex decays by transferring its energy and angular momentum to a quantum or a thermal reservoir, thus creating elementary excitations. The decay process at zero temperature has been studied numerically in Refs. [334,335], where conditions are given for the instability of the vortex line in terms of the strength of the interactions. The decay can also occur in a conserving way due to the nonlinear term in the GPE, which can redistribute energy among the modes in close analogy with a pressure gradient in classical fluids.

Vortex decay at finite temperature has also been considered [336]. In this case the driving dissipative force originates from the scattering of thermal excitations off the vortex line and the resulting vortex lifetime, which is estimated as the characteristic time for the vortex line to reach the condensate boundary, is similar to the damping rate of collective excitations in a BEC at finite temperature (see Sec. 3.3.4).

2.3.7 Multi-component GPE

Multi-component condensates are obtained by concurrent trapping and cooling of the same bosonic species, for instance ^{87}Rb in distinct hyperfine or spin levels or even of mixtures of two different species such as ^{87}Rb and ^{41}K. The extra degrees of freedom provided by the internal state variables give rise to genuinely new physics with no counterpart in single-component BEC's [86,337–345]. The multi-component nature of the order parameter can support new nonlinear coherent structures, such as topological vortices, monopoles, and various forms of solitons including topological ones (skyrmions).

Multi-component BEC's are described by a set of coupled GPE's, forming what is sometimes called a Vector GPE (VGPE). The N-species VGPE reads as follows,

$$-i\hbar\partial_t\Psi_a = -\frac{\hbar^2}{2m}\nabla^2\Psi_a + V_a(\mathbf{r})\Psi_a + \sum_{b=1}^{N} g_{ab}|\Psi_b|^2\Psi_a \qquad (72)$$

with $a = (1,\ldots,N)$, where the matrix g_{ab} describes the mutual s-wave scatterings between atoms in the various components. The numerical solution of the VGPE can be accomplished by relatively straightforward extensions of the techniques that are used for the single-component GPE [308,337,339,344]. However, the phenomenology of the solutions is vastly richer. For instance, Savage and Ruostekoski [340] looked for 3D ground states in cylindrical geometries by solving the imaginary-time VGPE with Runge-Kutta algorithms (see Fig. 29). They found skyrmionic solutions of the form

$$\begin{cases} \psi_+(\mathbf{r}) = \sqrt{n(\mathbf{r})}\left[\cos\alpha(\mathbf{r}) - i\sin\alpha(\mathbf{r})\cos\theta\right] \\ \psi_-(\mathbf{r}) = \sqrt{n(\mathbf{r})}\left[-i\sin\alpha(\mathbf{r})\sin\theta\exp(i\phi)\right] \end{cases} \qquad (73)$$

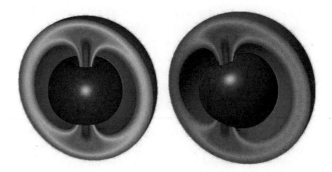

Fig. 29. 3D density and order parameter profiles for energetically stable trapped skyrmions. The central (blue) tori are isosurfaces of $|\psi_-|^2$; isosurfaces of $|\psi_+|^2$ (red to purple) are shown for $x < 0$. Left: stabilized by rotating ψ_- only with angular velocity 0.1ω. Right: stabilized by rotating the entire gas with angular velocity 0.085ω. From Savage and Ruostekoski [340].

where (α, θ, ϕ) can be understood as three spherical angles of a compact 3-sphere S^3 (a sphere in 4D). This S^3 sphere is topologically equivalent to the $SU(2)$ order-parameter space of a bi-spinorial condensate.

The topological stability of the skyrmions is controlled by their main topological invariant, which is the winding number

$$W = \frac{\epsilon_{ijk}}{24\pi^2} \int \mathrm{d}^3 r \mathrm{Tr}\left[U(\partial_i U^\dagger)U(\partial_j U^\dagger)U(\partial_k U^\dagger)\right] \tag{74}$$

where $U(\mathbf{r})$ is the 2×2 complex matrix mapping S^3 into $SU(2)$ parameter space and ϵ_{ijk} is the antisymmetric triple tensor. The main aim of these numerical studies is to assess the stability of the skyrmionic ground states against energetic perturbations such as shrinking to zero size, which can take place without changes in the winding number. In particular, numerical simulations offer means to inquire about possible stabilization mechanisms such as that provided by repulsive interactions between two components of a bi-condensate in the phase-separation regime, characterized by the condition $a_{++}a_{--} < a^2_{-+}$ on the scattering lengths. Proving stability in numerical simulations is not easy, since the basic requirement is that the solution be stationary in imaginary time and very long simulation times are needed. Similar studies have been made by Battye *et al.* [343], who employed conjugate-gradient routines for the direct minimization of the VGPE energy functional. Among other conclusions these authors confirm that an essential condition for skyrmion stability is phase separation of the components. This type of studies is opening up a fascinating new chapter in topological quantum-fluid dynamics.

Condensates belonging to two different hyperfine states which are coupled by a Raman transition provide an example of "internal" Josephson effect (see *e.g.* Sols [346]). Depending on the relative strength of the interactions the gas may be either in the Rabi regime where the atoms behave independently or in the Josephson regime where they display collective behaviour. The simplest model for the internal Josephson effect is a two-mode Hamiltonian

$$\mathcal{H} = -\frac{\hbar\omega_R}{2}(\hat{a}^\dagger\hat{b} + \hat{a}\hat{b}^\dagger) + \frac{\hbar\omega_C}{8}(\hat{a}^\dagger\hat{a} - \hat{b}^\dagger\hat{b})^2 \tag{75}$$

involving annihilation and creation operators for states A and B. In Eq. (75) ω_R is the Rabi frequency and ω_C is a "charge-frequency" defined by $4\hbar\omega_C = g_{AA} + g_{BB} - 2g_{AB}$. Kohler and Sols [341] find that in the presence of an effective attraction the particle losses spontaneously generate underdamped oscillations in the relative phase and number. With a decreasing number of atoms due to losses, the atom-atom interactions fade away and the gas leaves the bistable Josephson regime to enter a monostable Rabi regime characterized by oscillations in both phase and number of atoms. This study employs a quantum trajectory simulation method to handle the time evolution (for details see Sec. 3.3.5). The quantum simulations reveal that a classical description of the dynamics only applies in the high-loss regime, as is expected from analytical estimates.

2.3.8 BEC with attractive interactions

Condensates with attractive interactions collapse to a denser state as soon as the self-attractions overcome the zero-point motions induced by the confinement. Using the time-dependent GPE and simple extensions thereof, it has been possible to investigate the conditions for the stability of a condensate with attractive interactions, the dynamics of collapse, and the formation of bright solitons.

(i) Stability of attractive BEC under harmonic confinement. The mean-field attractions determine an absolute instability of the density distribution in a homogeneous Bose-condensed gas against a "collapsed" state corresponding to a liquid or a solid [347]. A confined BEC can instead exist as long as the self-interaction energy E_{int} per particle does not exceed the trap-level spacing ΔE. The maximum number of atoms allowed for a condensate in harmonic trap thus is

$$N_{max} \sim a_{ho}/|a| \tag{76}$$

since $\Delta E = \hbar\omega$ and $E_{int} \simeq gN/a_{ho}^3$ [348]. Early experiments on ^{7}Li have reported observation of a BEC with a limited number $N \simeq 1250$ of atoms [349]. Experiments on ^{85}Rb atoms have measured the constant $\kappa = N_{max}|a|/a_{ho}$ to be $\kappa = 0.46 \pm 0.06$ [350].

The solution of the static GPE and its dynamical evolution show a locally stable BEC only for small atom numbers [351,352] and yield an estimate for the constant κ which is not far from the experimental value, though systematically slightly larger [350,353]. A condensate state with $a < 0$ actually is a metastable state [354]: there also exists a collapsed state with a much smaller size, which can be reached through an energy barrier. The BEC state can thus decay through macroscopic quantum tunnelling [348,355–357] or by thermal fluctuations. Such processes are beyond the GPE framework.

(ii) Dynamics of condensate growth and collapse. When the atom number exceeds the maximum value the condensate collapses in a way similar to the decay of a star into a supernova [170]. In contrast to the stellar case, however, the BEC collapse stops when the density becomes sufficiently high to allow frequent three-body collisions, which cause a significant loss of atoms and bring the atom number back below threshold. Thus, in the presence of a reservoir the condensate can grow again and the process of condensate formation goes through a sequence of partial collapses in which the number of atoms in the condensate oscillates [358].

In an experiment on ^{7}Li gas the growth, collapse, and re-growth of the condensate have been observed, starting from a quenched cloud where most of the "hot" atoms are suddenly eliminated [359]. A second type of experiment addressing the dynamics of collapse has been performed with ^{85}Rb gas [360]. In this case the interactions among the atoms are suddenly turned from repulsive to attractive by driving them through a Feshbach resonance. As a result the condensate is observed first to shrink and implode, and then to emit bursts of high-energy atoms.

The collapse dynamics in atomic gases can be modelled with a GPE extended to include a three-body loss term, thus accounting properly for the early stages of collapse. That is, one solves the equation

$$i\hbar\,\partial_t\Phi(\mathbf{r},t) = \left[-\frac{\hbar^2}{2m}\nabla^2 + V_{ext}(\mathbf{r}) + g|\Phi(\mathbf{r},t)|^2 - i\frac{\hbar}{2}K_3|\Phi(\mathbf{r},t)|^4\right]\Phi(\mathbf{r},t) \quad (77)$$

where the three-body recombination rate K_3 for condensed atoms is six times smaller than for the thermal cloud (see Sec. 1.3.3). The value of K_3 varies strongly near a Feshbach resonance [361] and some details of the collapse process depend on its magnitude. Numerical simulations with small values of K_3 report a sequence of collapses and revivals of the condensate size [362], while larger values of K_3 lead to a single collapse event [358].

Equation (77) has been used to predict an oscillating dynamics during collapse as later observed in the ^{7}Li experiment. The experiment has also been simulated by using a quantum Boltzmann equation [359] (see Sec. 3.4.2). The use of Eq. (77) reproduces qualitatively the results of the ^{85}Rb experiment, although some small discrepancies are found in the estimate of the collapse time [353]. Thermal or quantum fluctuations, molecule formation or condensate fragmentation are possible candidates to explain the discrepancy between simulation and experiment.

(iii) Solitons. One-dimensional condensates with attractive interactions can form stable solitary waves propagating over large distances with no appreciable dispersion. The attractions compensate for the dispersion of the wavepacket and the condensate itself provides the required nonlinear medium. At variance from dark solitons (density dips) propagating in condensates with repulsive interactions, bright solitons (density spikes) occupy the whole condensate and are therefore extremely stable.

Bright solitons have been observed at Rice [201] and ENS [202] with ^{7}Li atoms in the hyperfine state $|F = 1, m_F = 1\rangle$. In this state, which can be trapped by optical means, the scattering length is positive [165] and allows the creation of a stable condensate. The scattering length is suddenly turned to negative by driving through a Feshbach resonance and the evolution of the condensate is observed under the effect of a weak antitrapping potential. At variance from the case of repulsive interactions, the condensate propagates without spreading. Formation of a soliton train has also been reported [201], with indications of a repulsion between the solitons.

The range of parameters for which soliton formation can occur in the ENS experiment has been estimated with a variational Ansatz on the GPE [202]. The soliton can exist only in a narrow window of parameters, since a short condensate is not enough to host stable 1D solitons while an elongated condensate tends to fall apart under the effect of the antitrapping potential. In addition, collapse will be induced if the strength of the nonlinearity is too large.

The formation of soliton trains as in the Rice experiment has been modelled with a variational Ansatz on the GPE [363] and by numerically solving the GPE with three-body losses [364] (see Fig. 30). The microscopic mechanism for soliton train formation is a modulation instability which can occur in condensates with attractive interactions [365]. In the quasi-homogeneous case the Bogoliubov dispersion relation $\hbar\omega_{\mathbf{k}} = \sqrt{(2gn + \xi_{\mathbf{k}})\xi_{\mathbf{k}}}$ becomes imaginary for $k < k_c = \sqrt{4|g|nm/\hbar^2}$, indicating the growth of an unstable mode of wavenumber $\bar{k} \sim k_c$. It turns out that the most unstable mode is $\bar{k} = k_c/\sqrt{2}$. The peaks in the soliton train are just the remnants of the density modulation induced by this mode, and the number N_s of solitons in the train can be estimated as $N_s \sim L/\bar{\lambda}$ by comparing the length L of the condensate with the wavelength $\bar{\lambda} = 2\pi/\bar{k}$ of the modulation [364].

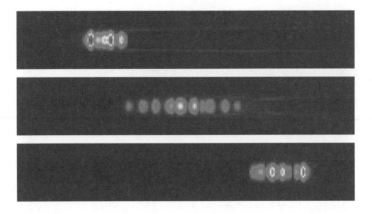

Fig. 30. An oscillating soliton train under harmonic confinement as obtained from a time-dependent GPE with losses using the parameters of the Rice experiment. From Salasnich *et al.* [364].

2.3.9 Dynamics of a 1D Bose gas in the Tonks limit

As already remarked in Sec. 2.2.4, a 1D Bose gas can be dilute and yet strongly interacting. The gas moves to the strong-coupling regime as its density is decreased and ultimately reaches the "fermionic" Tonks limit.

In a model where the density of the Tonks gas is determined by a wavefunction $\Phi(x, t)$ as $n(x, t) = |\Phi(x, t)|^2$, the appropriate energy functional is [223]

$$E[\Phi] = \int dx \left[\frac{\hbar^2}{2m} |\partial_x \Phi|^2 + V_{ext}(x) |\Phi|^2 + \frac{\pi^2 \hbar^2}{6m} |\Phi|^6 \right] . \qquad (78)$$

Eq. (78) contains as its self-interaction term the energy of a free Fermi gas, as prescribed by the boson-fermion mapping. The dynamics of the Tonks gas in a GPE-like mean-field approximation is obtained from Eq. (78) by functional differentiation ($i\hbar\partial_t \Phi = \delta E[\Phi]/\delta \Phi^*$). This leads to a nonlinear Schrödinger equation with a fifth-order term,

$$i\hbar \, \partial_t \Phi(x, t) = \left[-\frac{\hbar^2}{2m} \frac{\partial^2}{\partial x^2} + V_{ext}(x, t) + \frac{\pi^2 \hbar^2}{2m} |\Phi(x, t)|^4 \right] \Phi(x, t), \qquad (79)$$

where $V_{ext}(x, t)$ includes a time-dependent drive and the wavefunction $\Phi(x, t)$ is normalized to the number N of particles in the trap. This approach involving a single wavefunction is expected to be useful for small oscillations of the gas, although it overestimates its phase coherence since in a simulated interference experiment Eq. (79) would predict the formation of fringes with a sharper contrast than would appear in an exact solution of the propagation problem [366]. By construction Eq. (79) can also be used to describe the 1D ideal Fermi gas in a collisional regime, since for large values of N it has the same linear limit as the equations of hydrodynamics for fermions as long as one omits the surface-kinetic-energy contribution.

From the numerical solution of Eq. (79) one finds that the spectral frequencies are integer multiples of the trap frequency [367]. The same result is obtained by solving analytically the hydrodynamic equations in the Thomas-Fermi limit [367,368], even through the numerical solution of Eq. (79) shows that nonlinear terms in the kinetic energy density functional play a crucial role in determining the density profile near the Thomas-Fermi radius.

The spectrum of a 1D ideal Fermi gas under harmonic confinement in the collisionless regime is also given by the relation $\omega_n = n\omega_{ho}$. The dynamic structure factor $S(k, \omega)$ of this system can be evaluated exactly in terms of Hermite polynomials [369] (see Sec. 4.1). Figure 31 reports $S(k, \omega)$ for both the confined and the homogeneous gas at two values of the wavenumber, illustrating the profound changes that are induced by the confinement. It is remarkable that at such large numbers of atoms the spectrum under confinement is nevertheless well approximated by a local-density approximation on the structure factor of the homogeneous gas [369].

The measurement of the spectrum of collective modes of a quasi-1D Bose gas is expected to be a good probe for signatures of the Tonks-gas regime. For instance, the frequency of the quadrupole mode is $\omega = \sqrt{3}\omega_{ho}$ in the weak coupling limit [370–372] against $\omega = 2\omega_{ho}$ in the Tonks limit. The frequencies of the low-lying modes from weak to strong coupling have been numerically evaluated in Ref. [373].

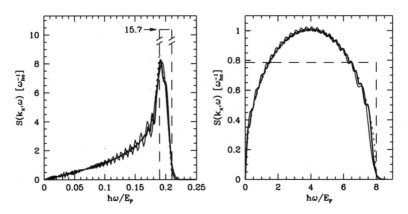

Fig. 31. Dynamic structure factor of a 1D harmonically confined Fermi gas as a function of $\hbar\omega/E_F$ where $E_F = N\hbar\omega_{ho}$. Left panel: at $k_x = 0.1\ k_F$, with numbers of fermions N=500 (solid line) and N=1000 (bold solid line). Right panel: at $k_x = 2\ k_F$ and N=20 (solid line) and N=40 (bold solid line). The spectrum of the homogeneous gas at the same values of the wavenumber is shown by dashed lines. From Vignolo *et al.* [369].

3 Bose gases at finite temperature

Thermal excitations develop in a Bose gas at increasing temperature and grow up to full depletion of the condensate. The critical temperature at which this occurs is affected by the strength of the interactions and the transition to a normal gas, being characterized by strong fluctuations, needs to be treated with methods beyond the mean-field level.

A two-fluid picture describes the gas as a "mixture" of a superfluid, phase-coherent component (the condensate) and a normal, dissipative component (the noncondensate), the relative population of the two components being dictated by thermodynamics. In coordinate space the thermal excitations tend to form a low-density halo around the condensate, and in momentum space they have a broad velocity distribution so that they contribute a substantial part of the internal energy of the gas. The collisions between thermal excitations and the condensate give rise to the damping of collective modes and to a decrease in the contrast of atom-laser output and of interference fringes.

The kinetics of the Bose gas at finite temperature is also very rich. Many fundamental questions have been investigated in parallel with the practical issues related to the cooling of the gas under various geometries, such as the onset of phase coherence and the growth of the condensate, the detailed balance between the populations of the two components and their collisions, and the mechanisms for the loss of superfluidity. A number of issues remain to be explored, like the kinetics in the proximity of the critical temperature.

A wide array of numerical methods has been developed to treat the Bose gas at finite temperature. These include classical and quantum Monte Carlo simulations, semiclassical and quantum kinetic approaches, finite-temperature extensions of the GPE framework, and functional integration methods for second-quantized many-body Hamiltonians. At variance from the zero-temperature case, where grid methods for the Gross-Pitaevskii equation and Diffusion Monte Carlo take the lion's share, the huge arsenal of numerical methods for finite-temperature BEC's is much harder to set out in a coherent framework. In this Section we shall provide an overview on the main techniques developed so far, leaving full details to the wide body of original publications.

3.1 Equilibrium properties

3.1.1 Two-fluid semiclassical model

A two-fluid model treating the equilibrium state of a bosonic cloud in terms of condensate and thermal-cloud densities was proposed in Ref. [374] in relation to the measurements by Ensher *et al.* [65] of the release energy in expanding clouds of ^{87}Rb atoms as a function of temperature (see Sec. 1.3.2). The model yields the typical bimodal atomic distribution that is used in the experiments as a signature of condensation (see Sec. 1.3.1) and is in fully quantitative agreement with the internal energy data shown in Fig. 4. It accounts for the observed deviation from the internal energy of the trapped ideal Bose gas below the critical temperature as due to the atom-atom interactions.

The two-fluid model is a mean-field theory using a semiclassical Hartree-Fock (HF) scheme for the thermal cloud and the Thomas-Fermi (TF) approximation to the GPE for the condensate to describe a dilute Bose gas with repulsive interactions. The TF approximation requires that the number N_c of atoms in the condensate be large ($N_c\, a/a_{ho} \gg 1$, see Sec. 2.1), while the HF scheme provides a simple and accurate description of the excitation spectrum in the presence of confinement [375], yielding results which are comparable with the predictions of the full Bogoliubov theory [66]. The semiclassical approximation requires that the thermal energy be appreciably larger than the harmonic-oscillator spacing and, as we shall see in Sec. 3.1.3, has been tested by a Path-Integral Monte Carlo study [376] for the parameters of the ^{87}Rb experiments.

The coupled equations for the condensate density $n_c(\mathbf{r})$ and the thermal-cloud density $\tilde{n}(\mathbf{r})$ are

$$n_c(\mathbf{r}) = \frac{1}{g}\left[\mu - V_{ext}(\mathbf{r}) - 2g\tilde{n}(\mathbf{r})\right]\theta(\mu - V_{ext}(\mathbf{r}) - 2g\tilde{n}(\mathbf{r})) \qquad (80)$$

and

$$\tilde{n}(\mathbf{r}) = \int \frac{d^3p}{(2\pi)^3}\left\{\exp\left[\frac{1}{k_BT}\left(\frac{p^2}{2m} + V_{eff}(\mathbf{r}) - \mu\right)\right] - 1\right\}^{-1}, \qquad (81)$$

where the effective potential acting on the thermal cloud is

$$V_{eff}(\mathbf{r}) = V_{ext}(\mathbf{r}) + 2gn_c(\mathbf{r}) + 2g\tilde{n}(\mathbf{r}) . \qquad (82)$$

In the above equations $g = 4\pi\hbar^2 a/m$ and the chemical potential μ is fixed by the normalization condition $\int d^3r\,[n_c(\mathbf{r}) + \tilde{n}(\mathbf{r})] = N$, the total number of atoms. The thermal component feels an effective repulsion from the interactions with the condensate atoms and is thus pushed as a thin cloud to the boundary regions, while the condensate forms a peak around the centre of the trap.

The two-fluid model has also been used to compute the specific heat and the momentum distribution of the gas at finite temperature [374,377]. Further applications have been to low-dimensional Bose gases [378] and to fermion-fermion and boson-fermion mixtures [379,171] (see Secs. 4.2 and 4.3).

3.1.2 Path-Integral Monte Carlo

A powerful approach to the thermodynamic properties of quantum many-body systems at finite temperature is the Path-Integral Monte Carlo (PIMC) technique [380]. PIMC methods provide a systematic way to compute matrix elements of the time evolution operator $U(\tau) = e^{-\mathcal{H}\tau/\hbar}$ in imaginary time. Upon interpreting imaginary time as an inverse temperature ($\tau/\hbar \leftrightarrow \beta = 1/k_B T$) and taking the trace of $U(\tau)$, one obtains the partition function $Z(T)$ of the quantum system at finite temperature T,

$$Z(T) = \text{Tr}\left(e^{-\beta\mathcal{H}}\right) = \int \left\langle \mathbf{R} | e^{-\beta\mathcal{H}} | \mathbf{R} \right\rangle \, d\mathbf{R} \tag{83}$$

where $\mathbf{R} = (\mathbf{r}_1, \ldots, \mathbf{r}_N)$ is the 3N-dimensional vector of the particle coordinates and $|\mathbf{R}\rangle$ the wavefunction in the coordinate representation. As it stands, Eq. (83) implies the calculation of the trace of an infinite-dimensional matrix, a task which can be performed analytically only for quadratic Hamiltonians. In the general case this trace must be evaluated by numerical means.

Upon discretizing the particle coordinates on a set of G grid points per spatial dimension, the computation of the trace amounts to performing a G^{3N}-dimensional integral, a task that can only be undertaken by Monte Carlo sampling. Numerical Monte Carlo consists in generating a sequence of configurations $\mathbf{R} \to \mathbf{R}' \to \mathbf{R}'' \ldots$ distributed according to the weight $\exp(-\beta\mathcal{H}[\mathbf{R}])$. This is equivalent to sampling the N-body Green's function $G(\mathbf{R}, \mathbf{S}; \beta) = \left\langle \mathbf{R} | e^{-\beta\mathcal{H}} | \mathbf{S} \right\rangle$, yielding the probability amplitude to move from the initial configuration $\mathbf{R}$ to the final configuration $\mathbf{S}$ in a "time lapse" $\tau = \hbar\beta$. Path-Integral Monte Carlo performs this task by generating a collection of paths connecting the end points $\mathbf{R}$ and $\mathbf{S}$. The partition function is then obtained on closing the paths by imposing $\mathbf{S} = \mathbf{R}$.

The main problem of this procedure is that the exponential operator appearing in the definition of the Green's function cannot be computed as a simple product of ordinary exponentials, since the kinetic energy operator $\hat{K}$ and the potential energy operator $\hat{V}$ do not commute. This problem can be circumvented by slicing the inverse temperature in M small intervals of size $d\tau = \hbar\beta/M$, so that $\exp(-\beta\mathcal{H})$ can be written as a direct product of M short-time terms, plus a correction of order $O(d\tau^2)$:

$$e^{-\beta\mathcal{H}} = \prod_{m=1}^{M} e^{K_m d\tau/\hbar} e^{V_m d\tau/\hbar} + O(d\tau^2). \tag{84}$$

In this so-called Trotter formula $\mathcal{H}_m = K_m + V_m$ denotes the local Hamiltonian at time $\tau_m = md\tau$. More explicitly,

$$G(\mathbf{R}, \mathbf{S}; \beta) = \prod_{i=0}^{M-1} \int d\mathbf{R}_i \langle \mathbf{R}_i | e^{-\mathcal{H}_{i+1} d\tau/\hbar} | \mathbf{R}_{i+1} \rangle \tag{85}$$

with $\mathbf{R}_0 = \mathbf{R}$ and $\mathbf{R}_M = \mathbf{S}$. As a result, numerical errors can be kept under control by choosing "small enough" values of $d\tau$. In passing, we note that low temperatures require correspondingly long simulation times.

The above description is correct for distinguishable particles. For indistinguishable bosons the correct symmetry of the N-body wavefunction under exchange must be taken into account and the appropriate expression for the generic state $|\mathbf{R}\rangle$ becomes $|\mathbf{R}\rangle = \sum_{\mathcal{P}} \mathcal{P} |\mathbf{r}_1, \ldots, \mathbf{r}_N\rangle / N!$ where the sum runs over all possible permutations of the particle positions. In applying periodic boundary conditions along the τ axis, the closed-path condition $\mathbf{S} = \mathbf{R}$ means that $\mathbf{S}$ should be matched to any of the possible permutations of all the initial configurations $\mathbf{R}$.

Permutations are increasingly more likely to occur on long paths at low temperature. In fact, while the identity is dominant at high temperature, all permutations become equally likely in the limit of zero temperature. The transition from the short-path to the long-path regime marks the onset of quantum degeneracy associated with the Bose-Einstein condensation transition. The superfluid fraction is linked to the long paths and can be estimated from the "winding number", which is an observable related to the "length" of the paths in permutation space. Special sampling techniques have been developed to preserve the efficiency of the Monte Carlo procedure in the degeneracy regime. A detailed discussion of these methods can be found in the review by Ceperley [380].

3.1.3   Equilibrium properties in harmonic traps

PIMC methods have been used by Krauth [381] to compute the density profiles and the condensate fraction as functions of temperature for $N = 10^4$ bosons interacting *via* a hard-core potential $V_{hc}(r)$ of radius a_0 inside a harmonic trap. The attainment of such a high number of bosons requires specific upgrades of the PIMC technique, aimed at increasing the time step at a given value of β. As discussed in Sec. 2.3.1 for explicit finite-difference methods, the number of bosons in the simulation sample sets a constraint on the upper allowed value of the time step.

In Krauth's work the N-body correlations are expressed as a product of N pair terms in the form

$$\rho(\mathbf{R}, \mathbf{S}; d\tau) = \prod_i \rho_1(i) \prod_{j>i} g_{12}(i,j) \tag{86}$$

where we have used the notation $\rho_1(i) \equiv \rho_1(\mathbf{r}_i, \mathbf{s}_i; d\tau)$ for the one-body density matrix and $g_{12}(i,j) \equiv \rho_2(\mathbf{r}_i, \mathbf{r}_j, \mathbf{s}_i, \mathbf{s}_j, d\tau)/[\rho_1(i)\rho_1(j)]$ for the two-body correlations. In a harmonic trap the one-body density matrix can be obtained

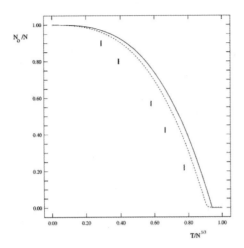

Fig. 32. Condensate fraction N_0/N in an isotropic trap *vs* reduced temperature $T/N^{1/3}$. The ideal-gas results are shown by a full line (in the thermodynamic limit $N \to \infty$) and by a dashed line (for $N = 10^4$). The symbols are PIMC results for 10^4 bosons interacting by a hard-core potential. From Krauth [381].

analytically and hence is easily sampled in an efficient manner using permutations. The two-body density matrix is also known analytically for hard-core interactions in the absence of confinement, and the effect of the confinement is included in an approximation using Gaussian factors. According to the author, this representation gains six orders of magnitude in efficiency as compared to a naive Trotter break-up.

The main findings of these calculations are as follows. The atom-atom interactions lead to a significant reduction of the condensate fraction N_0/N at finite temperature (see Fig. 32) and consequently to a lower critical temperature. The condensate wavefunction at the lowest temperature agrees with a simple GPE calculation to a high degree of precision and at finite temperature the thermal cloud forms a halo around the condensate in good agreement with the semiclassical two-fluid model review in Sec. 3.1.1.

The PIMC method has also been used to obtain more detailed information on a dilute Bose gas under harmonic confinement, such as its structure functions and equilibrium thermodynamics [382].

3.1.4   World-line Quantum Monte Carlo

The realization by Suzuki [383] that the partition function of a D-dimensional quantum spin-1/2 system can be mapped onto that of a $(D + 1)$-dimensional classical Ising model opened the way to quantum simulations on lattice models.

The "world-line" Quantum Monte Carlo (WLQMC) algorithm [258] is especially suited to lattice Hamiltonians. Its zero-temperature applications have been presented in Sec. 2.2.5, but since this method is a form of Path-Integral Monte Carlo some details are given below. Full details can be found in Ref. [292].

The world-line algorithm evaluates the partition function in the form

$$Z(T) = \sum_n \left\langle n | e^{-\beta \mathcal{H}} | n \right\rangle \tag{87}$$

where $|n\rangle = |n_1, n_2, \ldots, n_L\rangle$ is the wavefunction in the occupation-number representation, with $n_l = \langle a_l^+ a_l \rangle$ being the occupation number of the l-th site ($l = \{1, 2, \ldots, L\}$ for a system consisting of L lattice locations). The inverse temperature of the system plays again the role of an imaginary time. As for PIMC, the calculation of Eq. (87) requires handling huge matrices of size b^{L^D}, b being the number of single-particle states per lattice site. Such exponential escalation of degrees of freedom rules out standard diagonalization for all but the smallest lattices.

In a tight-binding approximation the WLQMC algorithm assumes that the Hamiltonian contains only first-neighbour terms and splits it into the sum over odd and even sites, $\mathcal{H} = \mathcal{H}_o + \mathcal{H}_e$. The reason for this checkerboard decomposition is that each term within the odd (even) Hamiltonian commutes with all other terms within the same Hamiltonian. On the other hand, the odd and even operators do not commute, so that a small-time Trotter-Suzuki decomposition is needed. To this purpose the imaginary-time direction is sliced into M units of time steps of size $d\tau = \hbar \beta / M$, so that one writes

$$e^{-\beta \mathcal{H}} \sim \prod_{m=1}^{M} e^{-(\mathcal{H}_{o,m} d\tau / \hbar)} e^{-(\mathcal{H}_{e,m} d\tau / \hbar)} + O(d\tau^2). \tag{88}$$

Since there are M exponentials in each component, it is natural to define a space-time lattice by stacking $2M$ lattices on top of the original lattice, so that each lattice site gets a double index (l, m) with $l = \{1, \ldots, L\}$ and $m = \{1, \ldots, 2M\}$. It is also convenient to associate the pair of indices (l, m) with a plaquette of area $d\tau dx$, whose south-west and north-east corners coincide with the sites (l, m) and $(l+1, m+1)$. It is easily checked that each local evolution operator $U_{lm} = e^{-\mathcal{H}_{lm} d\tau / \hbar}$ couples the four sites (l, m), $(l+1, m)$, $(l, m+1)$, $(l+1, m+1)$ around the plaquette, where (l, m) are either both odd or both even (shaded plaquettes in Fig. 33). Note that each $d\tau$ covers two values of the index m, one for the odd and one for the even terms, i.e. one time step is obtained by applying both components of the Hamiltonian.

This naturally induces a checkerboard decomposition of the space-time lattice such that interactions only take place on the corners of shaded plaquettes.

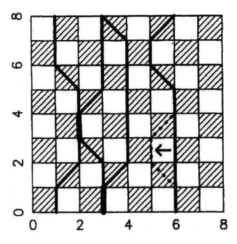

Fig. 33. The decomposition of the full tight-binding Hamiltonian into $\mathcal{H}_e + \mathcal{H}_o$ results in a checkerboard pattern for the space-imaginary time lattice. The dotted line illustrates a local QMC move. From Batrouni and Scalettar [384].

With these preparations everything is set to carry out the usual Metropolis Monte Carlo procedure: (i) change the occupation numbers $n_{l,m}$ by ± 1 ("spin-flip"); (ii) evaluate the resulting energy change δE, and accept/reject the move with probability $p = \min\left(e^{-\beta\delta E}, 1\right)$. One must keep account of the conservation law $n_{l,m} + n_{l+1,m} = n_{l,m+1} + n_{l+1,m+1}$ forbidding single "spin-flips", so that multi-particle moves must be devised. The simplest rule is to increase (decrease) by one unit the west corners of unshaded plaquettes and to decrease (increase) by one unit the east corners. This is tantamount to moving a particle right(left)-wards in real space, whence the "world-line" denomination of the algorithm. Note that particle number conservation prohibits breaking of the world-line. Global moves are also possible: for instance, one can follow a world-line which moves either vertically or diagonally across a shaded plaquette from top to bottom in a closed loop (see Fig. 33) and change the occupation numbers by ± 1 for each site along the world-line. These moves have often very low acceptance, but are necessary if the particles number (or the magnetization) is to fluctuate. Changes in the total number of particles require the adoption of a grand-canonical ensemble, with the addition of a term $-\mu \sum_l n_l$ to the Hamiltonian.

The world-line representation treats the diagonal terms of the Hamiltonian exactly and is therefore well suited to situations with diagonal-dominant Hamiltonians, as in bosonic Hubbard models. Its main advantage is that no sampling over permuted configurations is needed, since the second-quantized Hubbard Hamiltonian ensures built-in bosonic symmetry. On the other hand the presence of the lattice introduces further discretization errors besides the $O(d\tau^2)$

errors coming from the Trotter-Suzuki decomposition. Systematic errors arising from a finite time step can be virtually eliminated by the so-called continuous time technique [385].

Like most schemes based on lattice Hamiltonians, the world-line Monte Carlo method gives its best for 1D systems. To date, 1D WLQMC simulations can handle lattices with $L \sim 100$ sites and $N \sim 100$ bosons, as discussed in Sec. 2.2.5. However, 2D versions have also been developed and are currently used in numerical studies of high-T_c superconducting films and other strongly-correlated 2D quantum systems.

3.2   The BEC transition

Approaches beyond mean field are needed to obtain an understanding of the critical temperature and density at which Bose-Einstein condensation takes place. A review of theoretical studies of the phase transition has been given by Baym [227]. In 3D the transition temperature has been calculated by PIMC [386] and by the Worm algorithm [387]. In 2D condensation does not occur at any finite temperature in the homogeneous limit, but an interacting Bose gas is expected to undergo a transition of the Berezinskii-Kosterlitz-Thouless type [388,389] into a superfluid state (see Sec. 1.2.5). Present theories suggest that the critical density for this transition at a given temperature T is given by

$$n_c = \frac{mk_BT}{2\pi\hbar^2} \ln\left(\frac{\hbar^2\xi}{mU}\right) \qquad (89)$$

where U is the effective long-wavelength interaction and ξ is a dimensionless constant from an "ultraviolet" cutoff. The effective-interaction parameter is related to the zero-momentum Fourier component V_0 of the microscopic interaction potential by $U = V_0/(1 + V_0/V_c)$, where $V_c = (4\pi\hbar^2/m)/\ln(1/n_cd^2)$ with d the potential radius. In quasi-2D systems V_0 and d depend also on the confining geometry and on the 3D scattering length [14]. The specific value of ξ in Eq. (89) is beyond the reach of analytical methods, since it is controlled by the behaviour of the system in the region of strong fluctuations where perturbative expansions are inapplicable.

High-precision numerical methods are required for such studies. An adequate starting point for numerical strategies at small U is the classical $|\Psi|^4$ model with an effective long-wavelength Hamiltonian,

$$\mathcal{H}[\Psi] = \int \mathrm{d}^3x \left[\frac{\hbar^2}{2m}(\nabla\Psi)^2 + \frac{U}{2}|\Psi|^4 - \mu|\Psi|^2 \right]. \qquad (90)$$

The replacement of the quantum model by a classical one is justified in the critical region and brings about a substantial reduction of computational effort. The rationale behind the effective Hamiltonian in Eq. (90) is a clearcut separation between strong interactions at long wavelengths and irrelevant interactions at short wavelengths. The separation scale k_c in momentum space is defined as the inverse of the healing length, $k_c = 2\pi/r_c$ where r_c is the vortex-core radius [388, 389]. The complex field Ψ is then split into long-wavelength and short-wavelength fluctuating components, ψ and ϕ respectively. The dynamics of the slow component satisfies a Landau-Ginzburg-like equation with a stochastic forcing from short wavelengths. This Landau-Ginzburg equation is the Euler-Lagrange equation associated with the energy functional in Eq. (90).

The numerical solution of this type of stochastic Landau-Ginzburg equations is most conveniently handled by lattice Monte Carlo techniques. However, plain Monte Carlo faces an exponential inefficiency from critical slowing-down in the proximity of a phase transition. Critical slowing-down is due to the fact that the correlation length goes virtually to infinity as the critical density is approached, so that the generation of statistically independent Monte Carlo moves becomes exponentially inefficient. Powerful methods have been developed over the years to overcome this problem. In particular, cluster algorithms based on the idea of global moves have proved to be especially effective in curing critical slowing-down [390]. For instance, in the Ising model the growth of the correlation length is associated with the nucleation of large-size spin domains and the cluster algorithm moves all spins in the same domain at a time. However, BEC research has turned to a different and more recent class of high-precision Monte Carlo algorithms known as Worm algorithms, which we now review in some detail.

3.2.1 The Worm Monte Carlo algorithm

The price to pay for keeping up with critical slowing-down is the loss of locality - somewhat like we met in discussing implicit *versus* explicit finite-difference methods in Sec. 2.3.1. The special appeal of the Worm algorithm [391] is that (i) it manages to beat critical slowing-down without losing locality, and (ii) it provides direct estimators for the superfluid density, which is a major asset for the numerical study of a phase transition.

These remarkable results are obtained by exploiting some specific properties of closed-path (CP) configurations, produced by high-temperature expansions of lattice Hamiltonians. The key observation is that the configuration space of closed paths is sampled very efficiently through the motion of end points of disconnected paths - basically open world-line fragments named "worms". The Worm algorithm makes use of a number of ingenious and highly technical lattice combinatorics, whose details are best left to the original papers. Here we shall only set out some major guidelines, which are best illustrated on a simple Ising model.

(i) Partition function. The Ising Hamiltonian is first expressed as a sum over lattice bonds, $\mathcal{H} = \sum_{b=(i,j)} \mathcal{H}_b$ where the bond Hamiltonian is $\mathcal{H}_b = Ks_is_j$ with $K = J/k_BT$ being the coupling strength. Here s_i are the on-site spin variables and $b = (i,j)$ refers to a bond between nearest-neighbour sites on a square or cubic lattice. The partition function is then factorized into a product over the bond Hamiltonians,

$$Z = \sum_{\{s_i\}} \prod_b Z_b \tag{91}$$

where $Z_b = e^{-\mathcal{H}_b}$.

Each bond exponential is now expanded in a Taylor series of powers of the coupling strength K (high-temperature expansion),

$$Z_b[\{s_i\}] = \sum_{N_b=0}^{\infty} \frac{K^{N_b}}{N_b!} (s_is_j)^{N_b}. \tag{92}$$

The summation indices N_b are integer variables which can be attached to lattice bonds, thereby defining a so-called bond configuration $\{N_b\}$. For each bond configuration the summation defining Z can be factorized over the lattice sites,

$$Z = \prod_i Q(k_i) \tag{93}$$

where

$$Q(k_i) = \sum_{s_i=\pm 1} s_i^{k_i} \tag{94}$$

and $k_i = \sum_\nu N_{i,\nu}$ is the sum over bond states incident on node i. Here $N_{i,\nu}$ is an equivalent notation for N_b, whereby the bond $b = (i,j)$ is identified *via* the lattice site i and the direction ν to its nearest neighbour j. Each bond state can be graphically represented by N_b lines ("sticks") living on each lattice bond (see Fig. 34).

The CP constraint comes from symmetry: $Q(k_i)$ is non-zero only if the number of bond lines connected to site i is even. As an example, it is readily checked that for the Ising model $Q(k_i) = 0$ if k_i is odd and $Q(k_i) = 2$ if k_i is even. Each CP bond configuration contributes a weight

$$W_Z(\{N_b\}) = \left(\prod_b \frac{K^{N_b}}{N_b!} \right) \left(\prod_i Q(k_i) \right) \tag{95}$$

to the overall partition function,

$$Z = \sum_{N_b \in CP} W_Z(\{N_b\}). \tag{96}$$

90

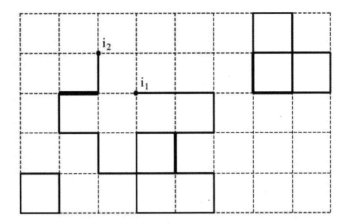

Fig. 34. An example of bond configuration for the correlation function $g(\mathbf{x},\mathbf{y})$, identified by indices i_1 and i_2. The width of the solid lines is proportional to the bond state number and lines with $N_b = 0$ are dashed. From Prokof'ev and Svistunov [391].

Thus, only sites connected with an even number of lines contribute to the partition function.

(ii) Spin-spin correlations. In a similar way one can face the core task, which is to evaluate the spin-spin correlation function $G\left(\mathbf{x}-\mathbf{y}\right) = g\left(\mathbf{x}-\mathbf{y}\right)/Z$ where

$$g(\mathbf{x}-\mathbf{y}) = \langle s(\mathbf{x})Z[s]s(\mathbf{y})\rangle \,, \tag{97}$$

with $[s]$ denoting a lattice spin configuration and the brackets standing for summation over all spin configurations.

Using the high-T expansion and the rearrangements described above, it is possible to show that the bond configurations contributing to spin-spin correlations differ from those contributing to the partition function in one respect only. Having chosen the sites $\mathbf{x}$ and $\mathbf{y}$, only paths with an odd number of lines can connect them. It is then natural to introduce the configuration space $\mathcal{C}_g$ for $g(\mathbf{x}-\mathbf{y})$, which consists of all bond configurations having an even number of bonds incident either on every lattice sites or on all but two sites. The key observation is that configurations for Z and for $g(\mathbf{x}-\mathbf{y})$ have identical bond elements since CP is an $\mathbf{x} = \mathbf{y}$ subset of $\mathcal{C}_g$. As a result, the non-normalized statistics for g and Z can be carried out within the same Monte Carlo sweep.

(iii) Worm algorithm. The Worm algorithm is built on the idea of sampling $\mathcal{C}_g$ through motions of end points only (open world lines). Visually, one may see the end point $\mathbf{y}$ as a magic pen which draws/erases a bond line when moving to a neighbour site. Here draw/erase means increase/decrease the bond number of the link by one unit.

The basic step consists of two updates for any two sites $\mathbf{x}$ and $\mathbf{y}$: (a) if $\mathbf{x} = \mathbf{y}$, move both with probability p_0 to a randomly chosen site $\mathbf{z}$ and shift $\mathbf{x}$ with probability $p_1 = 1 - p_0$ to a randomly chosen neighbour site; (b) else, shift $\mathbf{x}$ to a randomly chosen neighbour site. The overall Worm Monte Carlo procedure is then as follows:

(1) Starting from an arbitrary bond configuration, perform the Worm updates described above;
(2) The move and the shifts are accepted with probabilities determined by the set of functions $Q(k_i)$;
(3) Collect statistics if the updates are accepted and proceed to the next step.

(iv) Applications. The authors demonstrate the Worm algorithm for a variety of test cases, such as the 2D and 3D Ising model, the 2D and 3D XY model, and the 3D Gaussian model. In all cases but the $q = 3$ Potts model, they show impressive evidence for a logarithmic dependence of the correlation time on the system size,

$$\tau(L) = c_0 + c_1 \ln(L). \tag{98}$$

The $q = 3$ Potts model is found to scale as $\tau \sim L^{0.55}$, to be contrasted with $\tau \sim L^2$ of plain Metropolis Monte Carlo and $\tau \sim L^{0.515}$ of the Swendsen-Wang cluster algorithm.

The application of the Worm scheme to the $|\Psi|^4$ model is more elaborate, but still rather straightforward since the new terms as compared to the Ising model are purely local. In practice the only new ingredient is that for each bond there are two different terms, $\psi_i \psi_j^*$ and $\psi_i^* \psi_j$, which give rise to two separate expansions. This means that each bond state is defined by two numbers (N_b^1, N_b^2), which can be represented with a directed arrow.

3.2.2 Classical field simulations for phase transitions

The Worm Monte Carlo simulations allow one to obtain the critical density in a 2D Bose gas as a function of the system size L and of the coupling strength U [392,393]. The formal criterion of criticality is that the superfluid density n_s at the critical temperature has a universal value fixed by the Nelson-Kosterlitz relation [394],

$$n_s = \frac{2mk_BT}{\pi\hbar^2}. \tag{99}$$

The superfluid density is obtained from a direct Monte Carlo estimator in terms of winding numbers, namely the number of times a path ends on a periodic image of its starting point [34]. The simulations yield the value $\xi = 380 \pm 3$ for the numerical parameter entering Eq. (89).

A further important observable for a 2D Bose gas is the quasi-condensate density n_0 (see Sec. 1.2.5), which can be extracted from the simulation by means of a suitable two-body correlator [392]. The result at the critical temperature T_c is

$$\frac{n_0(T_c)}{n_c} \sim \frac{7.16}{5.94 - \ln\left(mU/\hbar^2\right)} \qquad (100)$$

where n_c is given by Eq. (89). The ratio in Eq. (100) is close to unity unless $mU/\hbar^2$ is exponentially small. The ratio of quasi-condensed to superfluid density is $n_0/n_s \sim 1.79$ at T_c, indicating that the superfluid density is significantly lower than the quasi-condensate density at the critical point.

3.3   Dynamics and transport

As we have previously seen, a consistent fraction of thermal non-condensed atoms appears when the temperature of the Bose gas is increased towards the critical condensation temperature. Experimental studies of the collective modes of a Bose-condensed gas at finite temperature have been reviewed in Sec. 1.3.2. Early calculations using the mean-field Hartree-Fock-Popov theory, which assumes a static thermal cloud, were found to be inadequate to explain some of experimental findings [395]. The dynamical coupling between the condensate and the thermal cloud is thus expected to play an important role except at the lowest temperature. Several theoretical approaches transcending a static mean-field approximation have been developed on the basis of second-order perturbation theory [396,397,81], a dielectric formalism [79,398], and a random phase approximation [399,400]. The latter treats the condensate and the thermal cloud on an equal footing and can be shown to reduce in the appropriate limits to second-order perturbation theory and to the dielectric formalism.

The inclusion of the dynamics of the thermal component in numerical studies also raises a considerable computational challenge. Several approaches have been developed for this problem, as we shall illustrate in the sequel.

3.3.1   Coupling the GPE with a Vlasov-Landau equation for the thermal cloud

An approach to deal with the dynamic evolution of both condensed and non-condensed components, in states where purely quantum fluctuations can still be neglected, is to couple a Gross-Pitaevskii equation for the condensate with a kinetic Vlasov-Landau equation (VLE) for the distribution function $f \equiv f(\mathbf{x}, \mathbf{p}; t)$ of the thermal excitations [401–403]. The coupled equations read

$$i\hbar\partial_t\Psi = \left[-\frac{\hbar^2}{2m}\nabla^2 + V_{ext} + g(n_c + 2\tilde{n}) - iR\right]\Psi\,, \tag{101}$$

$$\partial_t f + \frac{\mathbf{p}}{m}\cdot\nabla_{\mathbf{x}}f - \nabla_{\mathbf{x}}U\cdot\nabla_{\mathbf{p}}f = C_{12}[f] + C_{22}[f]\,. \tag{102}$$

In Eq. (101) $n_c(\mathbf{x},t) = |\Psi(\mathbf{x},t)|^2$ is the condensate density, $\tilde{n}(\mathbf{x},t) = h^{-3}\int d^3p$ $f(\mathbf{x},\mathbf{p};t)$ is the density of the noncondensate, and the term $R(\mathbf{x},t) = (4\pi h^2 n_c(\mathbf{x},t))^{-1}\int d^3p\,C_{12}[f]$ describes the transfer of atoms between the condensate and the noncondensate. In Eq. (102) the left-hand-side is the familiar streaming term in phase space, where $U(\mathbf{x},t) = V_{ext}(\mathbf{x},t)+2g(n_c(\mathbf{x},t)+\tilde{n}(\mathbf{x},t))$ is the effective potential acting on the noncondensate. Finally, $C_{12}[f]$ and $C_{22}[f]$ are condensate-noncondensate and noncondensate-noncondensate collision integrals, which reflect the quantum nature of the gas through their dependence on $(1 + f)$ enhancement factor from bosonic statistics.

Of course, several approximations lie behind Eqs. (101) and (102). As usual, a contact interatomic potential is used. More importantly, thermal excitations are treated in the Hartree-Fock approximation, i.e. via a self-consistent interaction term. The Popov approximation, namely the neglect of the anomalous density terms due to purely quantum fluctuations, is also made. The dilute-gas approximation will also be made in the collision integrals by appealing to two-body collisions and by neglecting correlations between the colliding particles.

3.3.2 Direct Simulation Monte Carlo

The numerical solution of the nonlinear integro-differential equations (101) and (102) calls for distinctive numerical techniques. To the GPE one can apply any of the various grid methods that we have previously discussed in Chapter 2. The VLE, like most kinetic equations living in six-dimensional phase space, is best handled by particle methods using Direct Simulation Monte Carlo (DSMC). The basic idea is to represent the distribution function $f(\mathbf{x},\mathbf{p};t)$ as a collection of N discrete "computational particles", which may not correspond to the atoms of the gas:

$$f(\mathbf{x},\mathbf{p};t) \sim f_N(\mathbf{x},\mathbf{p};t) = \sum_{i=1}^{N} W\left[\mathbf{x} - \mathbf{x}_i(t)\right]\delta\left(\mathbf{p} - \mathbf{p}_i(t)\right) \tag{103}$$

where W is a localized shape function interpolating the particle positions $\mathbf{x}_i(t)$ into the nodes of the fixed grid where the GPE is solved (see Figure 35).

The particle distribution is initialized by sampling particle positions and momenta with a Bose-Einstein local equilibrium distribution,

$$f(\mathbf{x},\mathbf{p};t=0) = \left[e^{\beta(U(\mathbf{x})-\mu+\mathbf{p}^2/2m)} - 1\right]^{-1}\,. \tag{104}$$

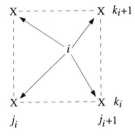 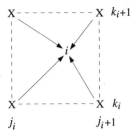

Fig. 35. Bilinear interpolation to weight (left) the contribution of particle i to the density at grid point (r_j, z_k) and (right) the effect of the forces at grid point (r_j, z_k) on particle i. From Vignolo *et al.* [403].

It is worth mentioning that this sampling is significantly more costly than for a classical gas, as the space and momentum variables do not separate. Sampling by inversion based on the Box-Müller algorithm [404] cannot be applied and an accept/reject sampling must instead be used. A plain accept/reject sampling becomes increasingly inefficient as the temperature is lowered, however, and some form of importance sampling must be devised to prevent the initialization stage from becoming a bottleneck.

Initialization is followed by the propagation step, which consists of a time-marching for the particle coordinates and momenta under the drive of the external and mean-field forces. An accurate choice is a slight variant of the Verlet method,

$$\begin{cases} \mathbf{x}_i(t + dt) = \mathbf{x}_i(t) + \mathbf{p}_i dt/m + \mathbf{F}_i(t) dt^2/2m \\ \mathbf{p}_i(t + dt) = \mathbf{p}_i(t) + (\mathbf{F}_i(t + dt) + \mathbf{F}_i(t)) dt/2 \end{cases} \tag{105}$$

where $\mathbf{F}_i = -\nabla U(\mathbf{x}_i)$. Besides being simple and accurate the Verlet algorithm is also symplectic, *i.e.* it preserves volumes in phase space in compliance of Liouville's theorem. This is an extremely valuable property, especially for very long-time integrations, where accumulation of small errors can lead to artificial dissipative effects.

Collisions are accounted for in a third step. For contact interactions a pair of particles can collide only if they are both in the same computational cell, and the probability for a collision to occur between a given pair inside a cell of volume dV in a time step dt is $p_{ij} = v_{ij}\sigma_{ij}dt/dV$, where v_{ij} is the relative speed of the particles and σ_{ij} is the scattering cross-section. Particle pairs are selected at random and tested for a collision by a standard accept/reject method, comparing with p_{ij} a random number in the range $[0, 1]$. Once a pair of particles is elected for a collision, the velocities are updated according to a scattering rule compliant with mass, momentum, and energy conservation.

This procedure is conceptually simple and computationally straightforward. It is also computationally efficient as long as the acceptance ratio can be kept reasonably close to unity. This hinges on the condition that the distribution of scattering probabilities be narrow.

The ideal case in this respect is provided by Maxwell molecules, whose interaction falls off with the inverse fifth power of separation yielding $v_{ij}\sigma_{ij} = const.$ so that all pairs have the same probability to collide. The cross-section is instead a constant in a bosonic thermal cloud, so that the probability for two particles to collide is inversely proportional to their relative speed. This implies that a certain rate of rejects in the collisional step cannot be avoided when the solution of the VLE is coupled to that of the GPE. Whenever dispersion in velocity becomes too wide, filtering techniques become instrumental to preserve acceptable efficiencies. Essentially the idea is that each particle i is tested for collision only with its fittest mate, namely the particle j which maximizes p_{ij}. This strategy will be discussed in more detail in Sec. 4.2.2 devoted to fermion mixtures, for which Monte Carlo simulation is made significantly more difficult by the need to comply with Pauli's exclusion principle.

All of the above refers to noncondensate-noncondensate collisions. Collisions leading to particle transfers between the condensate and noncondensate can be treated in a similar way. Thermal particles whose post-collisional energy falls below a threshold are transferred to the condensate population, while particles acquiring an energy higher than the threshold can "jump" into the thermal component.

3.3.3 Scissors modes in a condensate at finite temperature

Scissors modes are characteristic excitations of a Bose gas that are triggered by an angular displacement of an anisotropic trap (see Sec. 1.3.4). The theory predictes simple results for a condensate in the zero-temperature limit and for a gas above the condensation temperature. The intermediate temperature regime has been investigated experimentally by Maragó et al. [405] and studied theoretically by Nikuni [406] from the moments of the kinetic equations (101) and (102).

Jackson and Zaremba [401] have carried out a numerical study of the scissors modes at finite temperature within the GPE-VLE scheme, with specific attention to the observations of Maragó et al.. They solved the GPE by means of a spectral method and the VLE by Direct Simulation Monte Carlo, using ten computational particles per atom. Although the two components are only weakly coupled, the condensate oscillations experience a significant damping from the presence of the thermal cloud. The numerical results for the frequency shifts and the damping rates are in quantitative agreement with the

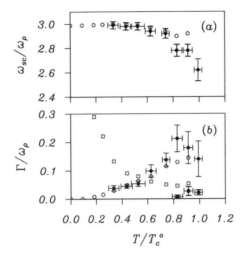

Fig. 36. Frequency and damping rate of the scissors modes for a choice of parameters as in the experiments of Maragó *et al.* [405]. Open symbols refer to numerical results and solid symbols to experimental data. The open squares in (b) show the calculated average damping rate of the two thermal-cloud modes, while the solid squares are the corresponding measured values. From Jackson and Zaremba [401].

experiment (see Fig. 36).

We cite at this point some other applications of the GPE-VLE approach. These have concerned the calculation of frequencies and damping rates for quadrupolar modes in a Bose gas inside a harmonic trap [80] and the dipolar oscillations of a gas driven through an optical lattice by a harmonic force [403]. The method has also been used to study the evaporative cooling of a Bose gas (see *e.g.* [407,408]) in alternative to the quantum trajectory method reviewed in Sec. 3.4.2.

3.3.4 Time-dependent projected GPE

As we have seen in Chapter 2, the GPE provides an excellent description of both equilibrium and dynamical properties of a BEC whenever quantum and thermal fluctuations can be neglected. It has been argued that GPE-like equations could also be used at finite temperature, the condition for applicability being that the thermally excited modes are sufficiently populated to justify a semiclassical treatment. In the limit $N_k \gg 1$ quantum fluctuations could be neglected ($\delta N_k / N_k \ll 1$), so that these modes could be described by a coherent wavefunction. The possibility of using an extended GPE at finite temperature deserves attention, as numerical procedures for its solution are well developed and significantly less expensive than those for semiclassical kinetic equations, let alone those for the fully quantal methods that will be reviewed in the following sections.

Of course the condition $N_k \gg 1$ is bound to break down with increasing energy, but this difficulty can be avoided by imposing an ultraviolet cut-off in momentum space. An elegant procedure for this has been developed by Davis et al. [409] for a homogeneous Bose gas. These authors set up a projection strategy, which eliminates the modes above a cut-off k_c defined by the condition $N_k \gg 1$ for $k < k_c$. This is done by introducing a projection operator $\hat{P}$ acting on the Bose field $\hat{\Psi}$ in a such a way that the projected field operator $\hat{\psi} \equiv \hat{P}\hat{\Psi}$ contains only modes with wavenumber below k_c. That is $\hat{\psi} = \sum_{|\mathbf{k}|<k_c} \phi_{\mathbf{k}}(\mathbf{x})\hat{a}_{\mathbf{k}}$ where $\hat{a}_{\mathbf{k}}$ are the usual boson annihilation operators and $\phi_{\mathbf{k}}(\mathbf{x})$ is a set of basis functions. The remainder $\hat{\eta} = (1 - \hat{P})\hat{\Psi}$ is a fluctuating, high-frequency field acting as a noise term in the time evaluation of $\hat{\psi}$.

In principle, this approach could form the starting point for a quantum thermal field theory [410]. As in any other kinetic theory, a closure is needed and requires an equation for the fast-fluctuating field $\hat{\eta}$. Setting $\hat{\eta} = 0$ yields a projected GPE for the field $\hat{\psi}$ and, since this equation preserves unitarity and reversibility, it is not clear how it could relax to thermal equilibrium. The rationale invoked by the authors, which is justified a posteriori by their numerical results, is an assumption of "microscopic chaos", namely that nonlinear, reversible, deterministic systems can display collisionless thermalization due to chaotic/ergodic behaviour in phase space [411]. This is an intriguing conjecture, that would deserve a study of its own since even its classical counterpart is far from being well established.

Davis et al. [409] give a numerical demonstration of the projected GPE on a homogeneus Bose gas. The calculations are carried out in k-space using spectral methods, which are ideally suited to the projection procedure. Starting from a randomized initial configuration and performing time marching by a fourth-order Runge-Kutta method with adaptive time step, their results reproduce equilibrium properties such as the fraction of condensed particles, as well as the Bogoliubov dispersion relation upon neglect of all coupling terms beyond second order.

3.3.5 Sampling methods for Wigner distribution

The GPE-Vlasov approach described in Sec. 3.3.1 is based on a semiclassical representation of thermal excitations. Yet another line of attack to a finite-temperature Bose gas faces both quantal and thermal fluctuations by resorting to a functional Wigner treatment of the bosonic field.

The principle of the method, which has been developed by Sinatra et al.

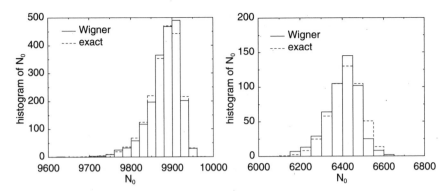

Fig. 37. Probability distribution in the canonical ensemble for the number of condensate particles in a 1D model (left) and a 2D model (right) of a harmonically trapped ideal Bose gas in thermal equilibrium at $k_B T = 30 \hbar \omega_{ho}$. The results of a truncated Wigner functional approach are compared with those of an exact Bogoliubov rejection method. From Sinatra et $al.$ [413].

[412,413], is to generate a Wigner functional and to sample it with an ensemble of classical fields evolving according to a generalized GPE. Sampling the Wigner functional generally is an ill-posed task because of lack of positive-definiteness. However, these difficulties can be overcome when the fraction of noncondensate particles is small, as the Gross-Pitaevskii operator reduces to a quadratic form so that the corresponding truncated Wigner (TW) functional regains positive-definiteness and standard numerical sampling procedures can safely be applied. As for the vast majority of functional equations, the numerical solution of the TW equation is handled by Monte Carlo techniques after discretization on a lattice of grid points. The corresponding phase space is then sampled for both equilibrium and non-equilibrium systems. The interested reader may find full details in the original publications.

The computational procedure has been tested on equilibrium properties of both ideal and interacting Bose gases [413]. The results for the ideal Bose gas in 1D and 2D are compared in Fig. 37 with exact results obtained by a Bogoliubov rejection method. The solution is very accurate and reproduces a marked deviation from Gaussian behaviour for the number of condensate particles inside a harmonic trap. This feature is found to persist in the case of an interacting Bose gas.

(i) Time-dependent Bose-condensed gas. In applications to time-dependent problems, phase space is sampled by stochastic realizations of the classical field $\psi(\mathbf{x}, t)$. A comparison is first made with the time-dependent Bogoliubov method, $i.e.$ a systematic expansion in powers of $\epsilon = 1/\sqrt{N}$ yielding the GPE in the limit $N \to \infty$.

More precisely, the field is written in the form

$$\psi = a_\phi \Phi + \psi_\perp \tag{106}$$

where Φ is the condensate wavefunction, $\psi_\perp$ is the noncondensate wavefunction, and a_ϕ is a normalization coefficient. Both Φ and $\psi_\perp$ are expanded in powers of ϵ, so that three types of coupling arise as a consequence of the expansion, i.e. $\Phi\Phi$, $\Phi\psi_\perp$, and $\psi_\perp\psi_\perp$. The first coupling describes a pure condensate with $(N-\delta N)$ particles, the third describes the dynamics of the noncondensate in the Bogoliubov approximation, and the cross coupling term describes the effects of particles in the noncondensate on the condensed ones as corrections to the GPE.

The TW equation is studied at fixed temperature T in the limit $N \to \infty$ at constant gN and for a fixed number $\mathcal{N}$ of components of the discretized field $\psi_\perp$. The system is found to relax to an effective classical temperature T_{cl} and comparison with the predictions of the Bogoliubov theory yields the conditions for validity of the TW approach as $N \gg \mathcal{N}/2$ and $(T_{cl} - T)/T \ll 1$. The first condition requires that on average there should be more than one boson per grid point or equivalently that the healing length should significantly exceed the grid spacing so that the grid "sees" a smooth particle distribution. This is a reasonable request for any explicit numerical scheme and should not be regarded as particularly restrictive.

The second condition is instead more peculiar and relates to the long-time dynamics of the gas. The authors show that T_{cl} is always larger than the actual temperature T of the system, the excess temperature being associated to quantum noise. In a regime where the excitations can be viewed as a collection of weakly coupled Bogoliubov oscillators, the quantum noise acts as a uniform source of energy for all modes and consequently results in classical equipartition on the set of oscillators. The error $(T_{cl} - T)/T$ scales like the square of the Bogoliubov energy, pointing to the TW approach being a nearly optimal stochastic method.

3.3.6 Stochastic wavefunction Monte Carlo

Another functional approach to the finite-temperature N-boson problem has been developed by Carusotto et al. [414,415]. The method bears strong similarities with the Wigner approach in that it makes use of stochastic field methods, but differs from it in a number of technical details. The authors present it as a new Quantum Monte Carlo method based on the stochastic evolution of classical fields rather than on walker trajectories of standard Path-Integral Monte Carlo. A distinct advantage is advocated for situations involving non-local observables.

The basic idea is to consider a stochastic GPE and to solve it for classical wavefunctions $\Phi_l(\mathbf{x}, t; N)$ $(l = 1, 2)$ representing a system of N atoms in the same quantum state. The set of stochastic differential equations obeyed by the wavefunctions in imaginary time reads

$$d\Phi_l = -\mathcal{H}_{GP}\Phi_l d\tau/2\hbar + idW_l \qquad (107)$$

in Ito form, where

$$\mathcal{H}_{GP} = \frac{p^2}{2m} + V_{ext} + g(N-1)\left(|\Phi_l|^2 - \frac{dV}{2}\sum_{\mathbf{x}'}|\Phi_l(\mathbf{x}')|^4\right) \qquad (108)$$

is a generalized Gross-Pitaevskii operator and dW_l is a stochastic noise having zero mean and variance proportional to $\sqrt{d\tau}$. In Eq. (108) dV is the volume of the grid cell. The Eqs. (107) are advanced in imaginary time up to the desired value of the inverse temperature.

The authors discuss two main choices for the wavefunctions, namely Fock (number) states and coherent wavepackets, and perform detailed studies of the indicators of statistical errors, typically the fluctuations of physically conserved quantities. Fine-tuning of the noise term, so as to minimize the statistical spread of the N-particle density matrix, yields much better stability (no finite-time blow-up of the solution) than so-called trace methods. Nevertheless, even the non-exploding schemes show an undesirable exponential growth of the statistical spread. Since the growth rate of these instabilities scales linearly with the number of bosons, the applicability of the method appears to be restricted to systems with a small number of particles.

The method has been applied to compute the distribution of condensate atoms as a function of temperature for a gas with $N = 125$ atoms [415]. The Monte Carlo simulations were made on a grid with $64 \div 128$ points using 3×10^4 realizations. The results are shown in Fig. 38.

3.4   Condensate formation and growth

Condensate growth has been experimentally studied by following the thermalization of the boson cloud after rapid ejection of hot atoms from the trap (see Fig. 39). The theoretical description of the formation and growth of a BEC has been tackled by a quantum kinetic treatments dealing with an open quantum dissipative system. These approaches are computationally expensive and are limited in the size of the cloud that they may handle, contrary to the case of the time-dependent GPE for which large-scale numerical simulations are possible and often extremely useful (see for instance Berloff and Svistunov [416].

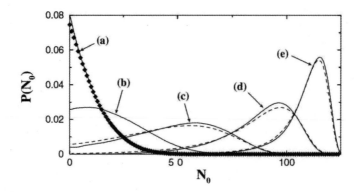

Fig. 38. Distribution function of condensate atoms at $k_B T/\hbar\omega_{ho} = 50$ (a), 33 (b), 20 (c), 10 (d), and 5 (e). The dashed lines are the Bogoliubov predictions, while diamonds refer to the ideal Bose gas. From Carusotto and Castin [415].

One of the chief tools of quantum kinetic theory are master equations for the density matrix operator. These can be derived from the Heisenberg equations of motion under the Markovian short-memory approximation [417], which is a canonical way to introduce dissipation and irreversibility within the quantum world. Dissipative quantum problems have received attention especially in the framework of quantum optics, a field from which BEC research has been borrowing a great deal of numerical techniques as we are going to see in the following.

3.4.1 Quantum master equation approach

A master-equation approach to the problem of condensate growth was proposed by Gardiner et al. [96] and further developed in several other papers [418–423]. In Ref. [96] the authors start by splitting the gas into two regions, the condensate band R_C and the noncondensate band R_{NC} (see Fig. 40). The region R_{NC} serves as a thermalized reservoir providing a supply of atoms for condensate growth. The region R_C includes both the condensate and most of the atoms whose dynamics is significantly affected by the presence of the condensate, in a set of energy levels up to a threshold E_R lying well above the condensate level. This region is treated fully quantum-mechanically in terms of atoms and quasiparticles, the latter being a superposition of phonon states whose dynamics is responsible for the transfer of atoms from excited states to the condensate level.

In this formulation the condensate is described by the band field operator

$$\hat{\Psi}_C(\mathbf{x}) = \hat{B}\left[\phi_N(\mathbf{x}) + \frac{1}{\sqrt{N}}\sum_m \left(f_m(\mathbf{x})\hat{b}_m + g_m(\mathbf{x})\hat{b}_m^\dagger\right)\right] \qquad (109)$$

102

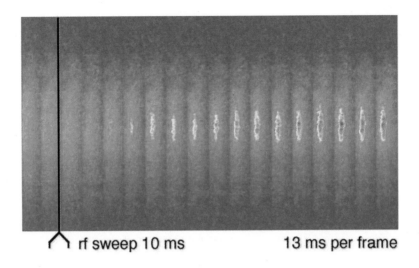

rf sweep 10 ms   13 ms per frame

Fig. 39. Observation of the formation of a ^{23}Na condensate. From Miesner *et al.*
[97].

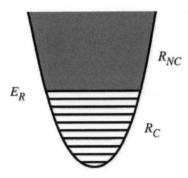

Fig. 40. Sketch of the condensate and noncondensate bands, from Gardiner *et al.*
[96].

where $\phi_N(\mathbf{x})$ is the condensate wavefunction obeying the time-independent
GPE, while $f_m(\mathbf{x})$ and $g_m(\mathbf{x})$ are the amplitudes for destruction/creation of
quasiparticles with energy ϵ_m at point $\mathbf{x}$. The operator $\hat{B}$ takes the R_C system
from the state with N atoms to the state with $N-1$ atoms. The basic picture
is that collisions in the noncondensate region will eventually move an atom
into the condensate region and conversely collisions between atoms in the two
regions will transfer an atom into R_{NC}. Quantum kinetic theory provides a
perfect framework to express this picture into a master equation for the density
operator describing the state of the condensate.

The master equation gives the time dependence of the occupation probabilities

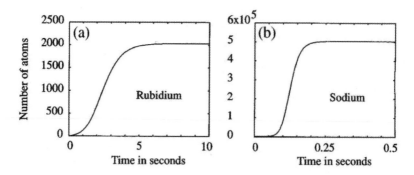

Fig. 41. Condensate growth for ^{87}Rb and ^{23}Na in a harmonic trap. From Gardiner *et al.* [96].

$p(\mathbf{n}, N)$, with $\mathbf{n} = \{n_m\}$ representing the whole set of quasiparticles in terms of the transition probability $W^{\pm}$ for atoms entering or leaving the condensate. A great simplification comes from the observation that transitions into the condensate are enhanced by a factor N from bosonic statistics, whereas cross-transitions only get a much smaller prefactor n_m. When all quasiparticles terms are accordingly dropped from the master equation, and on the assumption that the majority of the atoms are in the condensate, one can capture the essential behaviour of the master equation by simply examining the dynamical equation for the mean number n of atoms in the condensate. This equation reads

$$\frac{dn}{dt} = 2W^{+}(n) \left[1 + n\left(1 - e^{(\mu_n - \mu)/k_B T}\right)\right] \tag{110}$$

where μ_n and μ are the chemical potentials of the condensate and of the bath, respectively. Equation (110) describes the idealized limit in which the bath is not depleted as the condensate evolves, and for the specific case of a harmonic trap is easily solved numerically with the help of the Thomas-Fermi approximation for μ_n. As is shown in Fig. 41, the condensate growth follows a sigmoid evolution, with a latency period due to the accumulation of nonlinear effects followed by a rapid growth upon crossing a density threshold and finally by attainment of a steady state from nonlinear saturation of the growth instability.

In spite of the drastic simplification from the full-fledged master equation to the nonlinear differential equation (110), these numerical results are in semi-quantitative agreement with experiment [96,424–427]. The results in Fig. 41 are obtained with a constant bath temperature and chemical potential. A more realistic picture should include coupling of Eq. (110) with time-evolution equations for the evaporative cooling of the gas [428]. This more realistic model with various evaporative cooling schedules yields results in broad agreement with those shown in Fig. 41.

It thus appears that quasiparticle interactions do not affect the essential nature of the condensation process, which is basically captured by the simple growth equation (110). This may hinge on a general feature of a wide class of systems obeying "population-dynamics-like" evolutionary equations [429]. A specific example is offered by multi-mode lasers, in which the vast majority of the photons flow into the "fittest" (highest-gain) mode even when the gain is only marginally higher. A similar "winner takes all" behaviour has been observed in immune system dynamics [430].

3.4.2 Quantum trajectory methods

The master-equations approach leads to large systems of ordinary differential equations, which are very hard to solve if the number of participating modes is exceedingly large. A powerful alternative is offered by stochastic quantum trajectory methods. This class of quantum kinetic schemes is based on the idea of simulating the master equations by accumulating a number of stochastic realizations of wavefunction trajectories. In these methods the wavefunction evolves reversibly under the action of the relevant Hamiltonian and performs occasional "jumps" to a different quantum state. These jumps describe the non-unitary interaction of the system with a reservoir and provide the mechanism for dissipative effects. The appeal of these methods is that they permit to apply consolidated Monte Carlo practices to operator equations, which are otherwise hard to handle with deterministic methods.

The idea can be exported to the BEC framework, with the difference that quantum jumps no longer describe the coupling to a thermal reservoir, but rather the interatomic collisions in the gas. This idea has been developed into a concrete numerical scheme by Holland *et al.* [431].

(i) Quantum Boltzmann equation. Let us consider a dilute atomic gas in an isotropic harmonic trap of frequency ω. The Hamiltonian is the sum of a free term and a two-body term due to binary collisions,

$$\mathcal{H} = \sum_{\mathbf{n}} E_{\mathbf{n}} \hat{a}_{\mathbf{n}}^\dagger \hat{a}_{\mathbf{n}} + \sum_{\mathbf{n},\mathbf{m},\mathbf{p},\mathbf{q}} C(\mathbf{n},\mathbf{m},\mathbf{p},\mathbf{q}) \hat{a}_{\mathbf{n}}^\dagger \hat{a}_{\mathbf{m}}^\dagger \hat{a}_{\mathbf{q}} \hat{a}_{\mathbf{p}} . \tag{111}$$

Here $\mathbf{n} = (n_x, n_y, n_z)$, $E_{\mathbf{n}} = (n_x + n_y + n_z + 3/2)\hbar\omega$, and

$$C(\mathbf{n},\mathbf{m},\mathbf{p},\mathbf{q}) = \int \mathrm{d}^3x \, \mathrm{d}^3x' \phi_{\mathbf{n}}^*(\mathbf{x}) \phi_{\mathbf{m}}^*(\mathbf{x}') V(\mathbf{x},\mathbf{x}') \phi_{\mathbf{q}}(\mathbf{x}) \phi_{\mathbf{p}}(\mathbf{x}') \tag{112}$$

is the transition amplitude for the collision $(\mathbf{nm} \to \mathbf{pq})$.

In a dilute atomic gas the off-diagonal elements of the density matrix, $f_{\mathbf{nm}} = \mathrm{Tr}(\hat{\rho}\hat{a}_{\mathbf{n}}^\dagger \hat{a}_{\mathbf{m}})$ with $\hat{\rho}$ the N-body density matrix, can be adiabatically eliminated. The result is the Quantum Boltzmann equation (QBE) for the atomic populations $f_{\mathbf{n}} \equiv f_{\mathbf{nn}}$,

$$\partial_t f_{\mathbf{n}} = \sum_{\mathbf{m},\mathbf{p},\mathbf{q}} W(\mathbf{n},\mathbf{m};\mathbf{p},\mathbf{q})(f_{\mathbf{p}} f_{\mathbf{q}} \bar{f}_{\mathbf{n}} \bar{f}_{\mathbf{m}} - f_{\mathbf{n}} f_{\mathbf{m}} \bar{f}_{\mathbf{q}} \bar{f}_{\mathbf{p}}) \qquad (113)$$

where we have set $\bar{f} = 1 + f$. The transition rate is given by Fermi's golden rule,

$$W(\mathbf{n},\mathbf{m};\mathbf{p},\mathbf{q}) = \frac{2\pi}{\hbar}|C(\mathbf{n},\mathbf{m},\mathbf{p},\mathbf{q})|^2 \frac{\delta(E_{\mathbf{n}} + E_{\mathbf{m}} - E_{\mathbf{p}} - E_{\mathbf{q}})}{\hbar\omega}. \qquad (114)$$

The normalization condition is $\sum_{\mathbf{n}}^{N} f_{\mathbf{n}} = N_A$, where N_A is the number of atoms and N the number of eigenstates.

The QBE in Eq. (113) is a set of N nonlinear differential equations for the populations $f_{\mathbf{n}}$. A reduction in the complexity of the problem is achieved by assuming that the population within a single degenerate subspace is uniformly distributed among the degenerate states. This turns the vector quantum number $\mathbf{n}$ into a single scalar and cuts down the number of degrees of freedom by a cubic root. The reduced QBE keeps essentially the same form as Eq. (113), with a rescaling of the populations by means of the degeneracy factors $g_n = (n+1)(n+2)/2$. That is $f_{\mathbf{n}} \to g_n f_{e_n}$ where the index e_n runs over the non-degenerate energy levels.

Even with such a drastic simplification, the treatment of dynamical phenomena spanning several decades in temperature remains a computational challenge. The complexity of the calculation can be further reduced by using the classical limit of the QBE at all temperatures except those near the critical point. In Ref. [431] a smooth transition is found between the quantum and classical regimes, which permits to handle levels above a threshold as classical degrees of freedom. This transition smoothly takes the quantum factors g_n to their classical value $g_n = 1$, turning the quartic nonlinearity of the QBE into a quadratic one.

(ii) Kinetic evolution from simulated trajectories. The trajectory method to integrate the QBE starts by defining a trajectory function $f(e_n, t|e_\eta, t_\eta; \ldots; e_1, t_1)$, which describes a specific collision history ending up with energy e_n at time t. The trajectory is labelled by its history of η collisions occurring at times $t_1 < t_2 \cdots < t_\eta$ with energies $e_1, e_2 \ldots, e_\eta$.

The task of the numerical method is to produce a sequence of M trajectory realizations converging to the distribution $f_{e_n}(t)$ in the limit $M \to \infty$. The sum over these trajectories can be written as

$$g_n f_{e_n}(t) = \sum_{k=0}^{\infty} \sum_{e_1,\ldots,e_k} \int_{t_0}^{t} dt_k \int_{t_0}^{t_k} dt_{k-1} \cdots \int_{t_0}^{t_2} dt_1 f(e_n, t|e_k, t_k; \ldots; e_1, t_1),$$
$$(115)$$

106

where the sums run over sets of k possible collisions and all possible energies met before each of these collisions. These sums are then integrated over all possible times at which the collisions can occur. It can be shown that Eq. (115) converges to the solution of the original QBE. Note that each trajectory gets a uniform weight N/M, since there are M trajectories for N eigenstates. Realistic values of M in actual simulations are order $10^4 - 10^5$. The expression (115) is reminiscent of path-integration techniques, with the notable difference that time slicing is also changing with the actual realization of the stochastic process since the time lapse between two successive collisions is itself a stochastic variable. One may perhaps say that it is a form of path integration using continuous-time random walks [432].

The interested reader should refer to the original work for a detailed presentation of the computational procedure. Here we restrict ourselves to a few computational comments. The method offers a significant pay-off relative to the direct simulation of the full QBE. However, after making the assumption of isotropic occupation of the harmonic-oscillator states, the real computational savings come from replacing the QBE with its classical counterpart above an energy threshold. Indeed, in the classical case the statistical enhancement factors reduce to unity and the sums which define the collision operator can be performed analytically, thereby achieving linear complexity [428].

(iii) Results on condensate growth and evaporative cooling. In Ref. [431] the authors validate their method on a series of test cases and point out finite-number and finite-size effects as compared to the thermodynamic limit. The quantum trajectory method is ideally suited to study the build-up of the condensate in a dilute gas as well as the evaporative cooling of the gas all the way down to subcritical temperatures.

The condensate build-up is investigated by starting from a non-equilibrium distribution with no atom in the ground state and monitoring the ground-state occupation as the distribution evolves to equilibrium. Figure 42 shows the evolution of the ground-state population in time for a system of 100 bosons, of which about 10 percent enters the ground state at equilibrium.

The simulation of evaporative cooling is implemented by letting particles above a time-dependent energy threshold leave the system. In addition, atom trajectories can be lost by collisions with background atoms. Figure 43 shows the final numbers of atoms in the ground state as a function of the rate γ_{cut} at which the energy threshold $e_{cut}(t)$ changes exponentially with time. In line with physical intuition, for evaporative cooling to be successful the cut rate must be larger than the loss rate γ_{bl} to the background and lower than the initial single-particle collision rate $\gamma_{col}(t_1)$. An optimal cut rate can be found between these two limits.

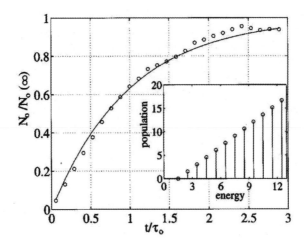

Fig. 42. Fraction of ground-state population, normalized to its time-asymptotic value. The inset shows the initial energy distribution (in units of $\hbar\omega$). From Holland *et al.* [431].

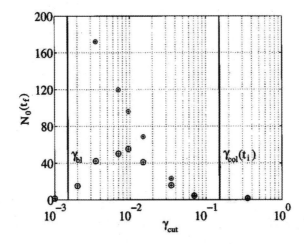

Fig. 43. Final number $N_0(t_f)$ of particles in the ground state as a function of the cut rate γ_{cut} with (circled crosses) and without (circled dots) background loss. The system consists of 10^4 bosons in thermal equilibrium at $T = 15T_c$. From Holland *et al.* [431].

Simulations of evaporative cooling with the QBE have been used to optimize the evaporative path (see *e.g.* [433]) and have been extended to study the cooling of boson-fermion and fermion-fermion mixtures [434–437].

4 Fermi gases and boson-fermion mixtures

The numerical study of fermionic matter is a most challenging task in computational physics. The difficulties all trace back to the need to obey the Pauli exclusion principle. In quantum field theory the anticommuting nature of the fermionic operators requires special mathematical objects known as Grassman numbers. These anticommute and, unlike ordinary real or complex numbers, have no direct representation on ordinary computers. In quantum many-body problems the need to preserve the antisymmetric nature of the many-body wavefunction under exchange of identical fermions raises again delicate numerical issues. Even a semi-classical representation of fermions must face the fact that a decreasing portion of phase space remains available for collisional events as the temperature is lowered. Such "Pauli blocking" implies that the task of performing fermionic collisions by means of Monte Carlo techniques meets with an exponential inefficiency barrier due to the statistical suppression factor from the occupancy of the final states. Here we shall be concerned mainly with the latter two types of difficulties.

The numerical evaluation of ground-state properties of a mesoscopic cloud of non-interacting fermions under harmonic confinement in various dimensionalities will be the first topic presented in this Chapter. We proceed to discuss interacting mixtures of Fermi gases and of bosonic and fermionic gases, whose theoretical study has received great impulse from experiments aimed at cooling fermions towards an expected superfluid state (see Sec. 1.4). The Chapter ends with a review of work on topics concerning strongly coupled Fermi gases and the crossover from a BCS to a BEC picture of fermionic superfluids.

4.1 Ideal Fermi gases under confinement

As already discussed in Sec. 1.4.1, a dilute Fermi gas occupying a single Zeeman sublevel of a magnetic trap can to a very good approximation be regarded as non-interacting since the Pauli principle suppresses s-wave collisions. Such a system is often referred to as a gas of spinless fermions. In the 1D case a precise mapping exists between spinless fermions and impenetrable bosons in the Tonks limit (see Sec. 2.2.4). Studies of low-dimensional Fermi gases under confinement also have an interest in connection with the physics of electrons in quantum dots (see *e.g.* [438,439]). We describe here the numerical methods which have been developed to treat an ideal Fermi gas under harmonic confinement.

The wavefunction $\psi_F(\mathbf{r}_1, \ldots, \mathbf{r}_N)$ for N spinless fermions at positions $\mathbf{r}_j$ in D dimensions is a Slater determinant of single-particle orbitals $\phi_i(\mathbf{r}_j)$ and the density profile is given by $n(\mathbf{r}) = \sum_i |\phi_i(\mathbf{r})|^2$. For harmonic confinement the orbitals are the solution of the Schrödinger equation for the D-dimensional harmonic oscillator, involving the Hermite polynomials and Gaussian factors. However, the use of Hermite polynomials has limited usefulness for numerical calculations on mesoscopic clouds, as discussed by March and Nieto [440]. Brack and van Zyl [441] have developed a powerful method for non-interacting fermions occupying a set of closed shells under isotropic harmonic confinement in D dimensions (see also [442]). They start from the expression of the one-body density matrix $\rho(\mathbf{r}, \mathbf{r}')$ as the inverse of the Laplace transform of the Bloch density matrix, leading to analytical expressions for the particle and kinetic-energy densities in terms of Laguerre polynomials $L_\alpha(x)$. For example in $D = 2$, when the Fermi energy E_F corresponds to $(M+1)$ filled oscillator shells, they obtain the expression

$$n(r) = \frac{2}{\pi a_{ho}^2} \sum_{\alpha=0}^{M} (-1)^\alpha (M + \alpha - 1) L_\alpha(2r^2/a_{ho}^2) \, e^{-r^2/a_{ho}^2} \qquad (116)$$

for the density as a function of the centre-of-mass position $r = |\mathbf{r} + \mathbf{r}'|/2$, and find

$$T(r) = \frac{\hbar\omega}{\pi a_{ho}^2} \sum_{\alpha=0}^{M} (-1)^\alpha (M + \alpha - 1)^2 L_\alpha(2r^2/a_{ho}^2) e^{-r^2/a_{ho}^2} \qquad (117)$$

for the kinetic energy density. Equations (116) and (117) are quite efficient for the purpose of numerical calculations. The same inverse Laplace method has been used to calculate the density profiles at finite temperature [443,444].

However, a Green's function method [445,446] is numerically more efficient in the case of strictly 1D confinement and for strongly anisotropic systems in $D \geq 2$. This method avoids the use of wavefunctions in favour of the matrix elements of the position and momentum operators, and can in principle be used for any confining potential and for systems at finite temperature. Of course, its application is especially simple in the case of harmonic confinement and has explicitly been shown to be practicable without any special numerical effort for mesoscopic clouds of up to 10^3 spinless fermions.

4.1.1   The Green's function method

Numerical techniques using Green's functions have been developed in solid state physics to evaluate the density of single-particle states and the transport properties of 1D systems of non-interacting electrons taken as models for polymers and quantum wires (see for example [447] and references therein). In view of the formal analogy between the density of states in the energy domain and the particle-density profile in coordinate space, one can introduce a Green's function operator $\hat{G}(x) = \lim_{\varepsilon \to 0}(x - \hat{x} + i\varepsilon)^{-1}$ where $\hat{x}$ is the 1D position operator and calculate the particle density at zero temperature as

$$n(x) = -\frac{1}{\pi} \lim_{\varepsilon \to 0^+} \text{Im} \sum_{i=1}^{N} \langle \phi_i | \hat{G}(x) | \phi_i \rangle \,. \tag{118}$$

A similar expression holds for all higher-order moments of the one-body density matrix, such as the kinetic energy density and its quantum fluctuations [448]. The method has been extended to evaluate the pair distribution function of the 1D Fermi gas, which determines its diffraction pattern [369]. We illustrate the evaluation of the particle density profile for spinless fermions at zero temperature in 1D harmonic confinement and the role of temperature in 1D and of trap anisotropy in 2D.

(i) Particle density profile in 1D. The trace of the operator $\hat{G}(x)$ on the first N quantum states determines $n(x)$ according to Eq. (118). In its evaluation one can make use of the decimation and renormalization procedures that have been introduced in the context of solid state theory [449]. The first step is to use a relation expressing the partial trace of a generic matrix through the determinant of its inverse matrix. The particle density may be written as

$$n(x) = -\frac{1}{\pi} \lim_{\varepsilon \to 0^+} \text{Im} \frac{\partial}{\partial \lambda} [\ln \det(x - \hat{x} + \lambda \mathbb{I}_N + i\varepsilon)] \Big|_{\lambda=0} \tag{119}$$

where $\mathbb{I}_N$ is a diagonal semi-infinite matrix with its first N eigenvalues equal to unity and null otherwise [445]. In the case of harmonic confinement the representation of the position operator on the basis of the eigenstates of the harmonic oscillator is a tridiagonal matrix and the calculation of the determinant in Eq. (119) can be conveniently carried out by a recursive algorithm. In practice the numerical evaluation of the determinant is performed through the product of its first M terms with $M \sim 10^7$. Figure 44 shows the particle density profile for various numbers of particles. The exact profiles are compared with those evaluated in the Thomas-Fermi approximation to emphasize the shell structure of the cloud and the spill-out of particles in the classically forbidden region (see the inset in Fig. 44(b)).

(ii) Extension to higher dimensionalities. For harmonic confinement the Green's function method has also been implemented in higher dimensionality [450] as for instance in $D = 2$ with an anisotropic confining potential. Since each wavefunction factorizes in the two Cartesian directions, the density profile can be written as

$$n_{E_F}^{2D}(x, y) = \sum_{i_x=1}^{I_x+1} \sum_{i_y=1}^{\tilde{I}_y+1} |\psi_{i_x}(x)|^2 |\psi_{i_y}(y)|^2 \,, \tag{120}$$

where the upper indices $(I_x, \tilde{I}_y)$ in the summation are fixed by the value of the Fermi energy E_F. The 2D density profile in Eq. (120) can be rewritten in the form of a 1D profile for $(I_x + 1)$ particles in the x direction as in Eq. (118),

111

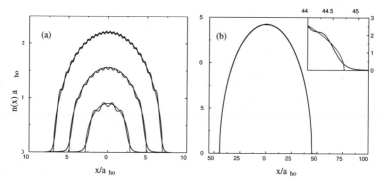

Fig. 44. Particle density profile as a function of position for harmonically confined spinless fermions or impenetrable bosons in 1D (bold lines), compared with the profile calculated in the Thomas-Fermi approximation: (a) for $N=4$, 12, and 24 particles; (b) for $N=1000$ particles, with an inset giving an enlarged view of the the spill-out region. From Vignolo al. [446].

but where the i_x-th term is weighted by the 1D profile of $(\tilde{I}_y + 1)$ particles in the y direction,

$$n_{E_F}^{2D}(x,y) = \sum_{i_x=1}^{I_x+1} \langle \psi_{i_x} | \delta(x - x_{i_x}) | \psi_{i_x} \rangle \, n_{\tilde{I}_y+1}^{1D}(y) \,. \qquad (121)$$

This expression allows a recursive application of the Green's function method.

Figure 45 shows the density profile for two closed-shell systems with about a thousand particles inside a trap with a large anisotropy, where only two or three levels are occupied in the y direction. The prominent shell structure which emerges in the direction of tight confinement reflects the macroscopic occupation of single quantum levels in this direction. An isotropic cloud with the same number of particles would show almost no shell effects and only display tiny oscillations in the density profile.

(iii) Extension to finite temperature. Considering for simplicity the 1D case [451], the particle density profile at finite temperature is the zeroth-order moment of the grand-canonical density matrix. This can be written in terms of the Green's function in coordinate space as

$$n(x) = -\frac{1}{\pi} \lim_{\varepsilon \to 0^+} \text{Im} \text{Tr} \left[\hat{T} \cdot \hat{G}(x) \right], \qquad (122)$$

where $\hat{T}$ is a diagonal matrix whose elements are the statistical Fermi factors.

The evaluation of the expression in Eq. (122) is carried out by a straightforward extension of the recursive procedure. Figure 46 shows how the shell structure in the density profile is washed away by thermal excitations.

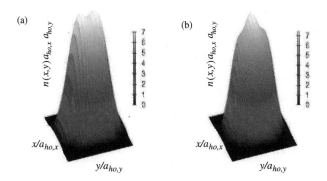

Fig. 45. 3D plot of the particle density as a function of $x/a_{ho,x}$ and $y/a_{ho,y}$ for (a) 1050 spinless fermions in a 2D harmonic trap with $E_F = 875\,\hbar\omega_x$ and anisotropy $k \equiv \omega_y/\omega_x = 350$, and (b) 1038 spinless fermions in a 2D harmonic trap with $E_F = 605\,\hbar\omega_x$ and $k = 173$. Redrawn from Vignolo and Minguzzi [450].

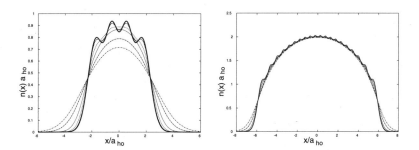

Fig. 46. Particle density profile for (a) $N = 4$ and (b) $N = 20$ harmonically confined spinless fermions in $D = 1$ at various values of the temperature: at $T = 0$ (solid curve) and $k_B T = 0.2\hbar\omega$ (dashed curve); the other curves refer to $k_B T/\hbar\omega = 0.5$, 1.0, 2.0 and 3.0, in order of decreasing peak height. Redrawn from Akdeniz et al. [451].

4.2 Dynamics of two-component Fermi gases

As already discussed in Sec. 1.4.2, experiments on colliding clouds of spin-polarized Fermi gases have shown that this setup can be an important tool to investigate the effects of Pauli blocking [166,173]. We report below on a numerical study of these phenomena.

A numerical approach which is often adopted to describe the dynamics of interacting Fermi gases is the so-called Fermi Molecular Dynamics (FMD). This method uses a quasi-classical scheme at the atomic and molecular level and has been developed in the framework of nuclear physics to describe the formation of antiprotonic atoms and more generally to treat nuclear structure [452,453]. Specific model potentials are used to constrain the fermionic dynamics to satisfy the Pauli exclusion principle in these high-density fluids of charged particles.

For dilute systems FMD is however inadequate and an approach explicitly exploiting the diluteness of the system is required. The first numerical effort that was directed to the transport properties of ultra-cold Fermi gases was reported in Ref. [454]. This work combines a finite-difference technique and a Monte Carlo method to solve the Vlasov-Landau equations (VLE) for the two-component fermionic Wigner distributions. The quantum-mechanical system is handled whithin a particle-dynamics approach, similarly to the dynamics of thermal bosons [403,455,456] and to FMD. In the following we will briefly review this approach, which embodies a treatment of a single spin-polarized Fermi gas [403] but supplements it with mean-field interactions and by collisions between the two components.

The distinctive mark of this numerical method is the strategy used to deal with collisional interactions. In particular, the development of a locally adaptive importance-sampling technique permits one to handle fermion-fermion collisions more rapidly by several orders of magnitude than in a naive application of Monte Carlo sampling. Collisions can be studied well below the Fermi temperature, down to $T \sim 0.2\, T_F$ where Pauli blocking would normally grind the simulation to a halt because of saturation of phase space and a vanishing efficiency of the Monte Carlo sampling.

4.2.1   Vlasov-Landau theoretical approach

The two fermionic components are denoted by the index $j = 1$ or 2 and are confined in external potentials $V^{(j)}(\mathbf{r})$. They are described by the distribution functions $f^{(j)}(\mathbf{r}, \mathbf{p}, t)$ which obey the kinetic VLE,

$$\partial_t f^{(j)} + \frac{\mathbf{p}}{m_j} \cdot \nabla_{\mathbf{r}} f^{(j)} - \nabla U^{(j)} \cdot \nabla_{\mathbf{p}} f^{(j)} = C_{12}\left[f^{(j)}\right] . \qquad (123)$$

The mean-field (Hartree-Fock) effective potential is $U^{(j)}(\mathbf{r}, t) = V^{(j)}(\mathbf{r}) + gn^{(\bar{j})}(\mathbf{r}, t)$, where $\bar{j}$ denotes the component different from j. We have set $\hbar = 1$ and $g = 2\pi a/m_r$, a being the s-wave scattering length between the two components and m_r their reduced mass, and $n^{(j)}(\mathbf{r}, t)$ are the spatial densities to be obtained by integration of $f^{(j)}$ over momentum. At equilibrium the distribution functions are given in a semi-classical Hartree-Fock approach as

$$f^{(j)}(\mathbf{r}, \mathbf{p}) = \left\{\exp\left[\beta\left(p^2/2m_j + U^{(j)}(\mathbf{r}) - \mu^{(j)}\right)\right] + 1\right\}^{-1}, \qquad (124)$$

where $\mu^{(j)}$ are the chemical potentials [457].

Collisions between atoms of the same spin are completely negligible at ultra-low temperatures, so that the collision integral C_{12} in Eq. (123) only involves collisions between particles with different spins,

$$C_{12}[f^{(j)}] = [2(2\pi)^4 g^2/V^3] \sum_{\mathbf{p_2,p_3,p_4}} \Delta_{\mathbf{p}}\Delta_\varepsilon \left[\bar{f}^{(j)}\bar{f}_2^{(\bar{j})} f_3^{(j)} f_4^{(\bar{j})} - f^{(j)} f_2^{(\bar{j})} \bar{f}_3^{(j)} \bar{f}_4^{(\bar{j})} \right]$$

(125)

with $f^{(j)} \equiv f^{(j)}(\mathbf{r},\mathbf{p},t)$, $\bar{f}^{(j)} \equiv 1 - f^{(j)}$, $f_i^{(j)} \equiv f^{(j)}(\mathbf{r},\mathbf{p}_i,t)$, and $\bar{f}_i^{(j)} \equiv 1 - f_i^{(j)}$. V is the volume occupied by the gas and the factors $\Delta_{\mathbf{p}}$ and Δ_ε are the usual delta functions accounting for conservation of momentum and energy, with the energies given by $p_i^2/2m_j + U^{(j)}(\mathbf{r},t)$.

4.2.2 Numerical solution method

The basic procedure to advance a VLE in time has already been reviewed in Sec. 3.3.2. In summary it consists of three steps: i) initialization of the fermionic distributions, ii) propagation in phase-space, and iii) inclusion of collisional interactions. Detailed accounts for Fermi gases can be found in Refs. [454,458].

i) The initial equilibrium distributions are generated starting with density profiles obtained from Eq. (124). Statistical noise is reduced by representing each fermion through N_q computational subparticles ("quarks"), with the restriction that no more than N_q quarks are allowed in any given cell of phase space.

ii) At each time step the quarks are driven by the external and mean-field forces in Eq. (123), using a standard time-marching scheme (typically a second-order symplectic integrator [459,402]). Even though no exclusion constraint is applied in this step, the distribution functions in Eq. (124), which fulfill the Pauli principle at all space points, evolve in time with no appreciable violations from the fermionic statistics if the propagation grid is sufficiently fine.

iii) Collisions are tracked on a coarser grid, the natural order being $vdt < dx < \delta x < l$ where v is a typical particle speed, dt the time step, dx and $\delta x \sim l_B$ the propagation and the collision mesh spacings. Here l_B is the de Broglie wavelength and l the particle mean free path. The leftmost inequality is dictated by accuracy and stability of the propagation step, whereas the rightmost one ensures accuracy and stability for the collision step.

The number of probable collisions between all possible pairs of quarks belonging to the two species in each cell of volume h^3 is evaluated at each computational time step as $dN_{coll} = dt \sum_{i,j} v_{ij}\sigma_{ij}$, where v_{ij} is the magnitude of the relative speed of particle i of species 1 and particle j of species 2, and σ_{ij} is the differential cross-section. The integer part of dN_{coll} gives the number of collisions and the remaining fractional part is interpreted as the probability of an additional collision. The pairs which collide are selected by a Monte Carlo sampling, the acceptance rate being enhanced by two-three orders of magnitude at each step by filtering out pairs with a classical collision probability below a dynamically adjusted threshold. The collision probability becomes smaller than the classical one after multiplication by the quantum suppression factor $1 - N(\mathbf{r}, \mathbf{p})/N_q$ due to the occupancy of the final state, and a further speed-up of the sampling is achieved by allowing each particle to collide only with the partner which maximizes the product $v_{ij}\sigma_{ij}$. This whole procedure permits one to address an otherwise unaccessible range of temperature.

4.2.3 Results for colliding fermion clouds

Figures 47 and 48 illustrate an application of the numerical method described above. After the experiments on colliding fermion clouds carried out at JILA, the dynamics of a two-component system representing magnetically trapped ^{40}K atoms in two hyperfine states $m_f = (9/2, 7/2)$ is being studied. The two components are axially displaced from the centre of the trap and start to oscillate with slightly different frequencies. In the absence of interactions the two clouds would keep oscillating at their respective trap frequencies without damping. The collisional rate Γ_q is varied by changing the mutual scattering length a and a transition from the collisionless to the collisional regime can be seen from the behaviour of the centres of mass of the two clouds (Fig. 47). The right panel of Fig. 47 shows the motion of the centres of mass when one of them is initially displaced while the other is left at the centre of the trap. This "kicked" run shows that at strong coupling the two clouds get locked together after only half a period of oscillation.

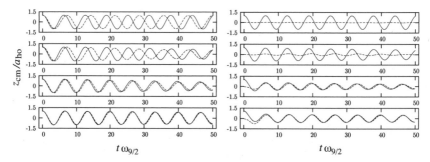

Fig. 47. Centre-of-mass oscillations for two fermion clouds as functions of time t (in units of the bare trap frequency $\omega_{9/2}$). From top to bottom the scattering length has the values $a = (150, 1500, 7500, 15000)$ Bohr radii. The two figures correspond to an initial axial displacement of both clouds (left) and of only one cloud (right). The simulation was performed at $T/T_F = 0.3$ with 2×10^4 fermions equally shared among the two spin states, each fermion being represented by ten quarks.

In Fig. 48 we plot the oscillation frequencies and the damping rates at $T = 0.3T_F$ as functions of the collision rate. At very low Γ_q the dipole frequencies are merely shifted away from the bare trap frequencies by the mean-field interactions. After an intermediate region where the data points show large fluctuations, the oscillations of the two clouds get locked together. In the collisionless region the damping rate varies linearly with Γ_q, while in the collisional one it scales like Γ_q^{-1} (right panel of Fig. 48). In fact, the hydrodynamic regime is reached at lower Γ_q as the temperature is lowered [454].

A further numerical application is reported in Fig. 49 and concerns the propagation of zero-sound waves in a non-collisional two-component Fermi gas as a function of the mean-field interactions [461]. The density profiles of an elongated cloud are perturbed by an effective potential simulating a laser beam continuously applied on the centre of the trap. The fermions are driven out and the density distortions move along the density profile with the zero-sound velocity $c_0 = \lambda v_F$, v_F being the Fermi velocity. While $\lambda = 1$ in the absence of mean-field interactions, it increases with the coupling strength. In Fig. 49 we compare the time evolution of the perturbed density profile in a non-interacting Fermi gas with those of a two-component interacting gas.

4.3 Confined boson-fermion mixtures

Dilute mixtures of bosons and fermions have been studied in several experiments by trapping and cooling atomic gases (see Sec. 1.4.1). The strength of the interactions can be varied by exploiting Feshbach resonances (see Sec. 1.4.5): the scattering length can be tuned on a wide range of values by means of a magnetic field [197,59]. Depending on the sign of the mutual scattering length two quantum phase transitions can occur at strong coupling (see Sec. 1.4.3): spatial demixing in the case of repulsive interactions, and collapse for attractive interactions.

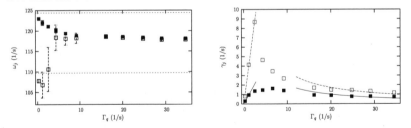

Fig. 48. Oscillation frequencies ω_j and damping rates γ_j as functions of the collision rate Γ_q for a mixture of ^{40}K atoms at $T = 0.3T_F$. The empty (filled) squares refer to the $m_f = 7/2$ ($m_f = 9/2$) component of the mixture. The horizontal dashed lines in the left panel give the bare frequencies of the two traps. Redrawn from Succi *et al.* [460].

The transition to spatial demixing can be identified by studying the equilibrium density profiles of the mixture [462–465] and affects dramatically the spectrum of collective modes, which shows a softening followed by a sharp frequency upturn [466,467]. Similarly, in the approach to collapse a softening of a family of collective modes is also found.

The dynamics of the mixture has been studied by semi-analytical methods using sum rules [468] or a variational Ansatz [469], as well as by numerical methods. In the collisional regime one solves numerically the equations of generalized hydrodynamics [466], while in the collisionless regime one uses the random-phase approximation (RPA) [467,470,471]. The latter is equivalent to a time-dependent Hartree-Fock approach. From a numerical point of view the hydrodynamic description becomes quite delicate in the demixing region, where the kinetic contributions to the tails of the density profiles cannot be neglected. On the other hand, the choice of a suitable Hartree-Fock basis is crucial to solve the RPA equations near collapse.

4.3.1 Boson-fermion mixtures at equilibrium

The Hamiltonian for a gaseous boson-fermion mixture can be written as

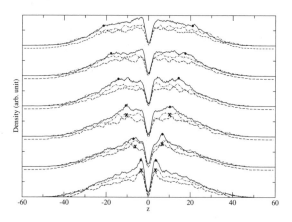

Fig. 49. Time evolution of a defect in the components of an interacting fermion mixture with $a = 2 \times 10^4$ Bohr radii (dashed lines), in comparison with a non-interacting Fermi gas (continuous lines). From bottom to top each profile corresponds to a time lag of 10 ms. Dots indicate the analytical estimate of the position of the peak in the non-interacting case and crosses indicate the peak positions corresponding to the zero-sound velocity. From Akdeniz et al. [461].

$$H = \sum_{\sigma=B,F} \int d^3r \, \hat{\Psi}_\sigma^\dagger \left(-\frac{\hbar^2 \nabla^2}{2m_\sigma} + V_{ext}^\sigma(\mathbf{r}) - \mu_\sigma \right) \hat{\Psi}_\sigma$$

$$+ \frac{g_{BB}}{2} \int d^3r \, \hat{\Psi}_B^\dagger \hat{\Psi}_B^\dagger \hat{\Psi}_B \hat{\Psi}_B + g_{BF} \int d^3r \, \hat{\Psi}_B^\dagger \hat{\Psi}_F^\dagger \hat{\Psi}_F \hat{\Psi}_B \tag{126}$$

where $\hat{\Psi}_\sigma$ are the bosonic and fermionic field operators, $V_{ext}^\sigma(\mathbf{r})$ are the confining potentials, μ_σ are the chemical potentials, and $g_{\sigma\sigma'}$ are the coupling constants expressed in terms of the s-wave scattering lengths $a_{\sigma\sigma'}$ as $g_{\sigma\sigma'} = 2\pi\hbar^2 a_{\sigma\sigma'}/m_{\sigma\sigma'}$, with $m_{\sigma\sigma'} = (1/m_\sigma + 1/m_{\sigma'})^{-1}$. Due to the Pauli principle there is no s-wave scattering between spin-polarized fermions ($a_{FF} = 0$). We focus on the case $a_{BB} > 0$ at zero temperature.

The mixture at equilibrium is described by a mean-field GPE for the boson condensate,

$$\left[-\frac{\hbar^2 \nabla^2}{2m_B} + V_{ext}^B(\mathbf{r}) + g_{BB} \, n_B(\mathbf{r}) + g_{BF} \, n_F(\mathbf{r}) \right] \Phi_B(\mathbf{r}) = \mu_B \Phi_B(\mathbf{r}) \tag{127}$$

and by a Schrödinger equation for the fermionic Hartree-Fock orbitals,

$$\left[-\frac{\hbar^2 \nabla^2}{2m_F} + V_{ext}^F(\mathbf{r}) + g_{BF} \, n_B(\mathbf{r}) \right] \psi_i(\mathbf{r}) = \varepsilon_i \psi_i(\mathbf{r}). \tag{128}$$

The particle densities are $n_B(\mathbf{r}) = |\Phi_B(\mathbf{r})|^2$ and $n_F(\mathbf{r}) = \sum_i |\psi_i(\mathbf{r})|^2 \theta(\mu_F - \varepsilon_i)$, and the chemical potentials are obtained by imposing the normalization conditions $N_\sigma = \int d^3r \, n_\sigma(\mathbf{r})$. At large particle numbers the fermion density is well reproduced by a Thomas-Fermi approximation supplemented by a surface kinetic energy term in the form of von Weizsäcker [472],

$$-\frac{\hbar^2}{6m_F} \frac{\nabla^2 \sqrt{n_F(\mathbf{r})}}{\sqrt{n_F(\mathbf{r})}} + An_F^{2/3}(\mathbf{r}) + V_{ext}^F(\mathbf{r}) + g_{BF} \, n_B(\mathbf{r}) = \mu_F \tag{129}$$

where $A = (\hbar^2/2m_F)(6\pi^2)^{2/3}$. The inclusion of the surface contribution is crucial in the solution of the hydrodynamic equations (see Sec. 4.3.3).

Let us consider first the case of repulsive boson-fermion interactions. As the value of a_{BF} increases the two species prefer to become localized in different regions of space. For the trapped mixture the overlap between the two clouds decreases over a broad range of coupling, as can be quantified by following the behaviour of the boson-fermion interaction energy at mean field level, $E_{int} = g_{BF} \int d^3r \, n_B(\mathbf{r})n_F(\mathbf{r})$. This quantity on increasing a_{BF} goes through a maximum and ultimately vanishes at "full demixing" (see Fig. 50).

In the case of attractive boson-fermion interactions the overlap between the two species increases with the strength of the coupling, until collapse occurs at

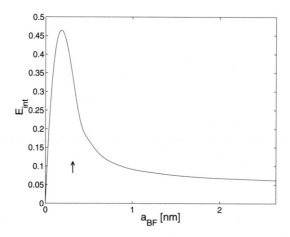

Fig. 50. Illustrating the boson-fermion interaction energy for a mixture of ^{6}Li-^{7}Li atoms in a trap as a function of the boson-fermion scattering length. The arrow marks the point where the fermion density vanishes at the trap centre. Redrawn from Capuzzi *et al.* [473].

the point where the attractions overcome the Fermi pressure and the boson-boson repulsions [462, 474-476].

In the homogeneous gas an analysis of the linear stability of the mean-field energy functional predicts the same condition for collapse and for demixing [477], that is

$$g_{BB}g_{FF} = g_{BF}^2 \tag{130}$$

with $g_{FF} = (2/3)An_F^{-1/3}$ playing the role of an effective fermion-fermion repulsion due to the Fermi pressure. A local-density approximation to Eq. (130) describes both the point of full demixing and of collapse for a boson-fermion cloud inside a trap.

4.3.2   Phase diagram and configurations in the demixed state

The numerical solution of the equilibrium equations yields several symmetry-breaking "exotic" configurations for the density profiles of the mixture in the demixed state. Figure 51 shows a phase diagram [478] summarizing the low-energy configurations indicated by letters and symbols for a ^{6}Li-^{7}Li mixture of experimental interest.

The plane $(a_{BF}/a_{BB}, k_F a_{BB})$ is divided into two regions by the curve given by Eq. (130). Above the demixing line a variety of different demixed configurations are found, three energetically stable structures being shown in Fig. 52. From an experimental point of view the column-density images corresponding to the configurations (b) and (c) may univocally identify the phase-separated regime [479].

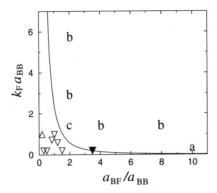

Fig. 51. Phase diagram of boson-fermion mixtures at $T = 0$ in the plane defined by the parameters a_{BF}/a_{BB} and $k_F a_{BB}$. The continuous line gives the condition of phase separation in Eq. (130). In the mixed-phase regime we have the following configurations; $\triangle$ fermions at the centre; $\triangledown$ bosons at the centre; $\blacktriangledown$ bosons at the centre in an almost demixed configuration. For the phase-separated regime we have used the following notations (see Fig. 52): (a) symmetric configuration with fermions outside; (b) symmetric configuration with fermions outside and in the central core; (c) boson torus inside a fermion cloud. Redrawn from Akdeniz et al. [464].

(a)　　　　　(b)　　　　　(c)

Fig. 52. Configurations of a trapped boson-fermion mixture in the phase-separated regime for $N_F = N_B = 10^4$ particles at $T = 0$. Redrawn from Akdeniz et al. [464].

4.3.3 Dynamics in the collisional regime

The hydrodinamic regime can be reached at high values of the scattering lengths, as already met in an experiment on a high temperature mixture [480]. In this regime the gas is best described by the equations of generalized hydrodynamics [481] for the particle density $n_\sigma(\mathbf{r}, t)$ and the current density $\mathbf{j}_\sigma(\mathbf{r}, t)$,

$$\partial_t n_\sigma + \nabla \cdot \mathbf{j}_\sigma = 0 \tag{131}$$

and

$$m_\sigma \, \partial_t \mathbf{j}_\sigma + \nabla \cdot \Pi^\sigma + n_\sigma \, \nabla \left(V_{ext}^\sigma + \sum_{\sigma'} g_{\sigma\sigma'} n_{\sigma'} \right) = 0 \,. \tag{132}$$

Under the assumption of high collisionality it is possible to express the kinetic stress tensor Π^σ in terms of the particle density and thereby one can close the hierarchy of dynamic equations. Consistently with Eqs. (127) and (129) for the equilibrium densities, Π^σ is given in a Thomas-Fermi approximation as

$$\Pi_{ij}^F = \frac{2}{5} A \, n_F^{5/3} \, \delta_{ij} - \frac{\hbar^2}{6 \, m_F} \left[\sqrt{n_F} \, \nabla_i \nabla_j \sqrt{n_F} - \nabla_i \sqrt{n_F} \, \nabla_j \sqrt{n_F} \right] \tag{133}$$

for the fermions and as

$$\Pi_{ij}^B = -\frac{\hbar^2}{2 \, m_B} \left[\sqrt{n_B} \, \nabla_i \nabla_j \sqrt{n_B} - \nabla_i \sqrt{n_B} \, \nabla_j \sqrt{n_B} \right] \tag{134}$$

for the bosons. In Eqs. (133) and (134) the velocity-dependent terms, which do not enter the linear dynamics, have been omitted.

The Fourier-transformed equations of linearized hydrodynamics have been solved in Ref. [466] for isotropic harmonic confinement by diagonalizing a set of eigenvalue equations in each subspace of fixed angular-momentum quantum number l. We report in Fig. 53 the low-lying monopolar ($l = 0$) frequencies as functions of the boson-fermion coupling in a ^{6}Li-^{7}Li mixture with $N_B \gg N_F$. We conventionally label each mode as "fermionic" or "bosonic" from the value of its frequency at zero coupling.

For positive a_{BF} (left panel in Fig. 53) the fermionic modes show a softening with increasing a_{BF}, followed by a sharp upturn in correspondence to the point where the fermion density at the centre of the trap vanishes [467]. This "demixing" point can be estimated as $g_{BF} = (\mu_F/\mu_B)g_{BB}$ and lies well before full demixing (see Fig. 50). The mode softening is a dynamical signature of the impending demixing transition, since on approaching the demixed state it costs less and less energy to excite monopolar oscillations. Consistently with this picture the density fluctuations of the two species become out-of-phase as a_{BF} is increased towards demixing. The softening does not continue down to zero frequency, however: as the topology of the fermion cloud changes at the demixing point, the boundary conditions for the hydrodynamic equations also change and this induces stiffening of the fermionic modes. In a mesoscopic system the transition to demixing can thus be sharply identified by looking at the mode frequencies.

For attractive boson-fermion interactions (right panel in Fig. 53) the fermionic modes exhibit a pronounced blue shift at intermediate values of a_{BF}. This is a consequence of the increase in the fermionic density as the strength of the interactions is increased: for $N_B \gg N_F$ the bosonic component, which is almost unaffected by the interactions, acts on the fermions as an effective attractive

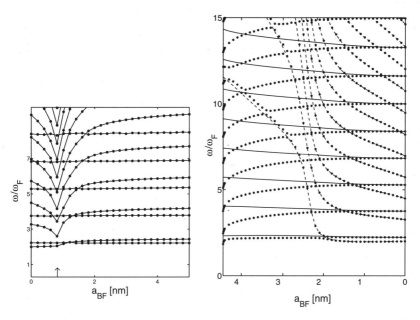

Fig. 53. Frequencies of $l = 0$ modes in a collisional boson-fermion mixture as functions of the scattering length a_{BF}. In the left panel the arrow indicates the point of vanishing fermion density at the centre of the trap and the lines are a guide to the eye. In the right panel solid and dashed lines are obtained by neglecting the dynamical coupling of the two species. Redrawn from Capuzzi *et al.* [466].

potential tightening the fermion trap. At larger values of $|a_{BF}|$ the bosonic modes show a softening, which becomes pronounced in the very proximity of collapse. The softening is an indication of an incipient collapse instability and also occurs in a pure condensate with attractive interactions [482,310].

4.3.4 Dynamics in the collisionless regime

Typical experiments on boson-fermion mixtures at the lowest feasible temperatures are carried out in the collisionless regime. However, many of the features predicted by the hydrodynamic equations are already relevant in regard to the experiments. For a dilute mixture the linear dynamics in the collisionless regime is well described by the RPA, which includes the dynamical coupling between bosonic and fermionic density fluctuations beyond mean field leading to mode repulsion and Landau damping.

The RPA assumes that the fluid responds as an ideal gas to the external perturbing fields $\delta U_\sigma(\mathbf{r})e^{i\omega t}$ plus the fields due to the Hartree-Fock fluctuations. The equations for the density fluctuations of the two species in Fourier

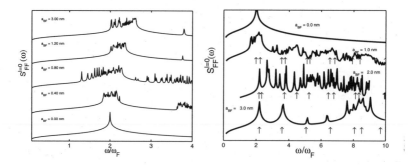

Fig. 54. RPA monopolar spectrum (in log scale and arbitrary units) for a boson-fermion mixture in harmonic confinement with $N_F = 10^4$ and $N_B = 10^6$, for various values of a_{BF}, positive in the left panel and negative in the right panel. The arrows in the right panel indicate the frequencies of the collisional modes from the right panel of Fig. 53. Redrawn from Capuzzi *et al.* [467].

transform with respect to the time variable are

$$\delta n_\sigma(\mathbf{r}, \omega) = \int d^3r' \, \chi^{0\sigma}(\mathbf{r}, \mathbf{r}', \omega) \left[\delta U_\sigma(\mathbf{r}') + \sum_{\sigma'} g_{\sigma\sigma'} \, \delta n_{\sigma'}(\mathbf{r}', \omega) \right]. \tag{135}$$

Consistency between equilibrium state and dynamics requires that the ideal-gas response functions $\chi^{0\sigma}(\mathbf{r}, \mathbf{r}', \omega)$ are built with inclusion of static mean-field contributions from solutions of the GPE and the Hartree-Fock equation. An example of the spectra excited by monopolar perturbing fields ($\delta U_\sigma(\mathbf{r}) \propto r^2$) can be found in Ref. [467] for a mixture in isotropic harmonic confinement. In this case the density fluctuations and the response functions can be expanded in spherical harmonics and one solves the dynamical equations for the monopolar modes in the $l = 0$ subspace.

The spectra of collective excitations are described by the dynamic structure factors

$$S_{\sigma\sigma'}(\omega) = \mathrm{Im} \int d^3r \, \delta n_\sigma(\mathbf{r}, \omega) \delta U_{\sigma'}(\mathbf{r}). \tag{136}$$

Figure 54 shows the fermionic monopolar spectrum in the low-frequency range for repulsive (left panel) and for attractive (right panel) boson-fermion coupling. At zero coupling this spectrum shows a single fermionic mode at frequency $2\omega_F$. A bosonic mode lies at frequency $\sqrt{5}\omega_F$ (for $\omega_B = \omega_F$), but does not appear in the fermionic spectrum in the absence of coupling. The monopolar fermionic mode becomes fragmented as its degeneracy is lifted by the coupling with the bosons. This phenomenon is also found in the static Hartree-Fock spectrum and is an intrinsic property of the Fermi gas, which occupies a large number of single-particle levels inside the Fermi sphere.

For $a_{BF} > 0$ a group of fermionic modes shows softening followed by an upturn, and this transition occurs again at the point where the topology of the fermion

124

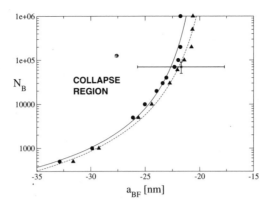

Fig. 55. Phase diagram of the ^{87}Rb-^{40}K mixture at $T = 0$. The data points indicate the boundary for collapse from the Thomas-Fermi theory with $N_F = 10^4$ (circles) and $N_F = 2 \times 10^4$ (triangles). The data point with error bars shows the collapse point observed in the Florence experiment [175]. Redrawn from Capuzzi *et al.* [483].

cloud changes by the opening of a central hole.

For $a_{BF} < 0$ the spectrum becomes first very fragmented and then shows only bosonic peaks at low frequency and mostly fermionic peaks at relatively high frequency. Comparison with the modes in the collisional regime (Fig. 53, right) suggests that the above behaviour of the spectrum can be related to the rapid blue shift of the fermionic modes, which first leads to several mode crossings and ultimately drives the fermionic modes to high frequency.

4.3.5   The case of the ^{87}Rb-^{40}K mixture

In the ^{87}Rb-^{40}K mixture the boson-fermion scattering length is large and negative, and collapse has been experimentally observed (see Sec. 1.4.3). In the experiment the mixture has been driven towards collapse by progressively increasing the number of bosons.

The equilibrium properties of the mixture have been calculated at the semi-classical Hartree-Fock level both at $T = 0$ [176,483] and at finite temperature [484]. The phase diagram at $T = 0$ is in good agreement with the measured location of the collapse (see Fig. 55), even though the calculations have neglected the experimental trap anisoptropies and misalignments. The location of collapse is very sensitive to the value of a_{BF} and allows its determination [176].

The dynamical spectra of the mixture have been studied in both the collisional and the collisionless regime [483], choosing for simplicity an isotropic harmonic trap. As the number of bosons is increased the bosonic collective modes in the

125

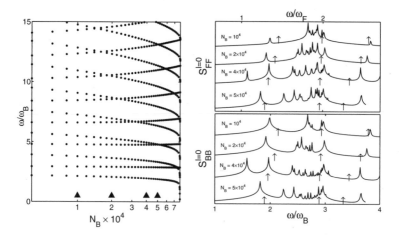

Fig. 56. Monopolar frequencies and RPA spectral response for a ^{87}Rb-^{40}K mixture with $N_F = 2 \times 10^4$ and varying N_B. Left panel: mode frequencies in the collisional regime (the triangles indicate to the values of N_B for which the RPA calculations in the right panel have been performed). Right panel: RPA fermionic and bosonic spectra, in log scale and arbitrary units (the arrows indicate the frequencies of the collisional modes for the same parameters). Redrawn from Capuzzi et al. [485].

hydrodynamic regime show a softening, which is most pronounced in the very proximity of collapse (left panel of Fig. 56). The softening is still small in magnitude for the parameters used in the RPA calculations (right panel of Fig. 56). At variance from the ^{6}Li-^{7}Li mixture in Figs. 53 and 54 the fermionic collective modes show only a modest blue shift, since for the experimental set of parameters the number of bosons is only slightly larger than that of fermions.

4.4 Fermi superfluidity and BCS-BEC crossover

Two very different extreme regimes can be vizualized for superfluidity in a two-component Fermi gas (see Sec. 1.4.4). The first is the BCS regime, where the normal state is a degenerate Fermi liquid that undergoes a pairing instability at a temperature $T_c = T_{BCS}$ such that $k_B T_c \ll E_g$ with E_g the ground-state energy of the non-interacting Fermi gas. The formation of Cooper pairs and their condensation occur simultaneously at the transition. The second regime is a BEC of bosons which are composite objects made up of an even number of fermions. These bosons condense at a temperature $T_c \ll T_{diss}$, with T_{diss} being a dissociation temperature.

In this section we review the models and the numerical studies that have

126

allowed estimates of the critical temperature and give access to the equilibrium and the dynamical properties of the Fermi gas with attractive interactions. We consider both the BCS limit and the crossover from a superfluid phase of weakly bound fermions to Bose-Einstein condensation of strongly bound composite bosons. The formation of a pseudogap in a high-T_c regime will also be discussed.

4.4.1 The critical temperature

An illustrative interpolation scheme for the critical temperature in the crossover from BCS to BEC has been proposed in a functional integral formulation to allow for an attractive coupling between fermions of different spins [184]. An effective action $S[\Delta]$ is obtained from the Hamiltonian density as a function of a Hubbard-Stratonovich field Δ which couples to the product $\hat{\Psi}_\downarrow^\dagger(\mathbf{r})\hat{\Psi}_\uparrow^\dagger(\mathbf{r})$ of field operators for the two spin species. After regularizing the ultraviolet divergence originating from the use of a contact potential, the gap equation is obtained from the condition $\partial S/\partial \Delta = 0$. The critical temperature is determined from the solution of the equation

$$\frac{m}{4\pi\hbar^2 a} = \frac{1}{V}\sum_\mathbf{k}\left[\frac{1}{2\epsilon_\mathbf{k}} - \frac{\tanh(\beta_c(\epsilon_\mathbf{k} - \mu)/2)}{2(\epsilon_\mathbf{k} - \mu)}\right], \tag{137}$$

where a is the scattering length, V is the volume, and $\epsilon_\mathbf{k} = \hbar^2 k^2/2m$. Equation (137) must be solved together with the number equation $n = n_0(\mu, T) + \delta n_{corr}(\mu, T)$ where $n_0(\mu, T)$ is the ideal Fermi distribution and $\delta n_{corr}(\mu, T)$ includes the effects of bound pairs in the strong-coupling normal state by allowing for Gaussian fluctuations.

In the weak-coupling limit $(1/a \to -\infty)$ the solution is essentially unaffected by the inclusion of fluctuations and one recovers the BCS result $T_c \simeq T_{BCS} \propto T_F \exp(-\pi/2k_F|a|)$. In the opposite limit $(1/a \to \infty)$ the fluctuations play a crucial role and the critical temperature asympotically tends to the value $T_c = 0.218\, T_F$. The same results have been obtained by Perali et al. [486] within a diagrammatic scheme (see also Sec. 4.4.4). The calculated critical temperature is reported in Fig. 57 as a function of the parameter $(k_F a)^{-1}$ which drives the crossover from BCS to BEC.

It has been proposed that a considerable increase of the BCS critical temperature of the homogeneous gas could be obtained by confining the fermionic atoms in an optical lattice. A study based on Path-Integral Monte Carlo simulations of the fermionic Hubbard model [487] has been carried out by Hofstetter et al. [488]. At sufficiently low temperatures and in a situation such that there is about one atom per lattice site, the fermions are confined to the lowest

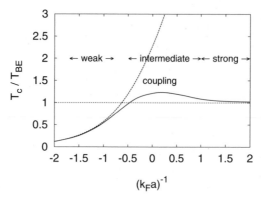

Fig. 57. Critical temperature T_c (full line) and BCS mean-field critical temperature (broken line) as functions of $(k_F a)^{-1}$. Both temperatures are normalized to the Bose-Einstein condensation temperature T_{BE} evaluated at the same density. Redrawn from Perali et al. [486].

energy band and can be described by the negative-U Hubbard Hamiltonian

$$\mathcal{H} = -t \sum_{\{i,j\},\sigma} (\hat{a}_{i\sigma}^{\dagger}\hat{a}_{j\sigma} + \hat{a}_{j\sigma}^{\dagger}\hat{a}_{i\sigma}) + U \sum_{i} \hat{n}_{i\uparrow}\hat{n}_{i\downarrow} \tag{138}$$

where $\hat{a}_{i\sigma}$ and $\hat{a}_{i\sigma}^{\dagger}$ are destruction and creation operators for a localized fermion of spin σ on site i and $\hat{n}_{i\sigma} = \hat{a}_{i\sigma}^{\dagger}\hat{a}_{i\sigma}$ (see also Sec. 2.2.5). In such a model the BCS-BEC crossover is governed by the parameter $|t/U|$: the limit of large tunnelling ($|t/U| \gg 1$) corresponds to the BCS weak-coupling limit, while strong localization ($|t/U| \ll 1$) coincides with the strong-coupling BEC regime. In experiments the crossover could be controlled by varying the depth U_0 of the lattice potential. The highest critical temperature would be achieved in the intermediate region where the interaction and tunnelling energies are comparable and atoms start forming pairs on single lattice sites. In the strong-coupling regime the large well depth would limit the mobility of the composite bosons and hence reduce T_c. The behaviour of the critical temperature as a function of the lattice-well depth is shown in Fig. 58.

Near a Feshbach resonance a Fermi gas with attractive interactions can be described by a coupled fermion-boson model [185,490–493],

$$\mathcal{H} = \sum_{\mathbf{k},\sigma} \epsilon_{\mathbf{k}} \hat{a}_{\mathbf{k}\sigma}^{\dagger}\hat{a}_{\mathbf{k}\sigma} + \sum_{\mathbf{k}} (E_{\mathbf{k}}^0 + 2\nu)\hat{b}_{\mathbf{k}}^{\dagger}\hat{b}_{\mathbf{k}}$$
$$+ g \sum_{\mathbf{q},\mathbf{k},\mathbf{k}'} \hat{a}_{\mathbf{q}/2+\mathbf{k}\uparrow}^{\dagger}\hat{a}_{\mathbf{q}/2-\mathbf{k}\downarrow}^{\dagger}\hat{a}_{\mathbf{q}/2-\mathbf{k}'\downarrow}\hat{a}_{\mathbf{q}/2+\mathbf{k}'\uparrow} \tag{139}$$
$$+ g_r \sum_{\mathbf{q},\mathbf{k}} \left(\hat{b}_{\mathbf{q}}^{\dagger}\hat{a}_{\mathbf{q}/2-\mathbf{k}\downarrow}\hat{a}_{\mathbf{q}/2+\mathbf{k}\uparrow} + h.c. \right) .$$

Here $\hat{a}_{\mathbf{k}\sigma}^{\dagger}$ is the creation operator of a Fermi atom with kinetic energy $\epsilon_{\mathbf{k}} = \hbar^2 k^2/2m$, $\hat{b}_{\mathbf{k}}^{\dagger}$ is the creation operator of a quasi-molecular boson with kinetic

128

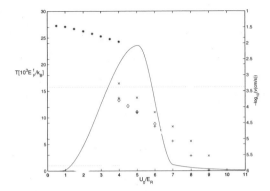

Fig. 58. The transition temperature for ^{6}Li atoms in a lattice as a function of the lattice-well depth. The symbols show the amplitude of Bloch oscillations (right-hand scale). Redrawn from Rodriguez and Törmä [489].

energy $E_{\mathbf{k}}^0 = \hbar^2 k^2/4m$, and ν is the threshold energy of the resonance. The Hamiltonian in Eq. (139) includes not only a standard interaction with coupling constant g, but also a coupling with strength g_r between fermionic monomers and composite bosons, as in a model introduced by Ranninger and Robaszkiewicz [494].

The superfluid transition temperature has been determined by Ohashi and Griffin [493] using the Thouless criterion, which associates the onset of the transition to the appearance of a singularity in the four-point particle-particle scattering vertex function related to the formation of Cooper pairs. The resulting equation for T_c can be viewed as a generalization of the gap equation (137) in the presence of a Feshbach resonance and reads

$$1 = \left(-g + \frac{g_r^2}{2\nu - 2\mu}\right) \frac{1}{V} \sum_{\mathbf{k}} \frac{\tanh(\beta(\epsilon_{\mathbf{k}} - \mu)/2)}{2(\epsilon_{\mathbf{k}} - \mu)}. \qquad (140)$$

Equation (140) is regularized by introducing a momentum cut-off and is solved together with the number equation $N = N_F + 2N_B + 2N_C$ for the numbers of free fermions, of bosons, and of Cooper pairs. Figure 59 shows the critical temperature and the chemical potential in the BCS-BEC crossover for a harmonically confined Fermi gas as obtained from Eq. (140) using a local density approximation for the chemical potential.

Quite remarkably, in the BEC limit the critical temperature for the harmonically trapped Fermi gas is predicted to be $T_c \simeq 0.518\,T_F$, which is twice that of the homogeneous gas.

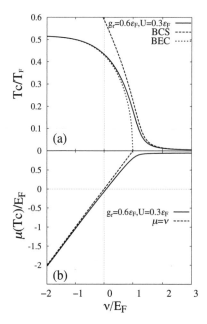

Fig. 59. BCS-BEC crossover in a harmonically trapped Fermi gas: (a) superfluid transition temperature T_c and (b) chemical potential μ at T_c as functions of ν. In this figure BCS labels the results in the absence of fluctuations, and BEC refers to a Bose-condensed gas of $N/2$ molecules of mass $2m$. Redrawn from Ohashi and Griffin [493].

4.4.2 Equilibrium properties

Calculations of the equilibrium properties for a Fermi gas in the BCS-BEC crossover have concerned the magnitude of the gap and the equilibrium particle density profile and momentum distribution.

(i) Gap at $T = 0$. The gap Δ has an analytic expression at mean-field level for the homogeneous 3D system at zero temperature, both in the BCS and in the BEC limit [495,496]. The BCS gap is exponentially small and reads $\Delta \simeq E_F \exp\left(-\pi/(2k_F|a|)\right)$. In the BEC limit one has $\Delta \simeq [16/(3\pi)]^{1/2}|E_F^3\mu|^{1/4}$ with $\mu = -E_b/2$, $E_b = \hbar^2/ma^2$ being the pair binding energy.

The gap and the ground-state energy have been numerically evaluated in the intermediate regime by a fixed node Green's function Monte Carlo (GFMC) simulation [497]. The GFMC (for a general reference [498]) is a variant of DMC (see Sec. 2.2.1). In an application of GFMC or of any other Monte Carlo approach to Fermi systems, odd permutations must get a negative sign in the transition from a configuration vector $\mathbf{R}$ to another. This is the fermion sign problem, which can be handled to some extent by the fixed-node approximation [499]. Here one solves the Schrödinger equation with the boundary

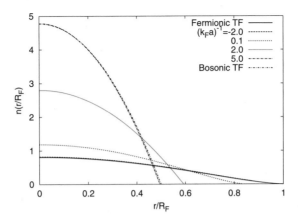

Fig. 60. Density profile in an attractive Fermi gas as a function of the radial coordinate for several values of the coupling parameter $(k_F a)^{-1}$. Redrawn from Perali et al. [193].

condition that the many-body wavefunction vanishes when a trial wavefunction does. This gives the best upper bounds consistent with the assumed nodes. The upper bounds on the ground-state energy and on the gap have been calculated to be $E_g = 0.44\,E_F$ and $\Delta = 0.81\,E_F$, using a Jastrow-BCS trial wavefunction to include pair correlations.

(ii) Density profile in a trapped gas. The zero-temperature density profile of a two-component Fermi gas with attractive interactions has been evaluated in Ref. [193] as a function of the coupling strength. The superfluid state is described by a generalized Thomas-Fermi approximation which covers continuously all coupling regimes, a local-density approximation being used for the fermionic chemical potential entering the gap equation (137). Figure 60 shows the density profile in an isotropic trap for various values of the parameter $(k_F a)^{-1}$. In the BCS limit the profile agrees with that of an ideal Fermi gas, while in the strong coupling limit $((k_F a)^{-1} = 5$ in Fig. 60) it agrees with that of a condensate of bosons with mass $2m$. An enhancement of the central density in the BEC regime has been observed in the experiments (see Sec. 1.4.5). The appearance of a peak in the density at the centre of the trap also is a signature of resonance superfluidity in the model of Eq. (139) [492].

Effects of increasing temperature on the density profile have been studied in the BCS regime by Houbiers *et al.* [178], using again a local-density approach to the gap equation.

(iii) Momentum distribution. The functional integral approach has been used in studying the momentum distribution of the homogeneous gas in the BCS-

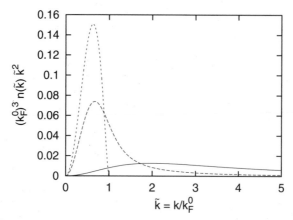

Fig. 61. Momentum distribution of a trapped Fermi gas in the BEC (solid line), intermediate (long-dashed line), and BCS (short-dashed line) regimes. From Viverit *et al.* [501].

BEC crossover [500]. At $T = 0$ one finds

$$n(\mathbf{k}) = \frac{1}{2V}\left(1 - \frac{\epsilon_\mathbf{k} - \mu}{\sqrt{(\epsilon_\mathbf{k} - \mu)^2 + \Delta^2}}\right) \tag{141}$$

in the BCS weak-coupling limit, while in the BEC limit one can write [501]

$$n(\mathbf{k}) = \frac{4}{3\pi V}(k_F a)^3 \frac{1}{(k^2 a^2 + 1)^2}. \tag{142}$$

This result is from the Fourier transform of a molecular wavefunction $\phi(r) \propto \exp(-r/|a|)/r$.

The momentum distribution of a homogeneous gas has been calculated at $T = 0$ in the intermediate regime by a fixed node GFMC simulation [497]. For a cloud under harmonic confinement a local-density estimate on Eqs. (141) and (142) has been made [501]. The profiles for isotropic confinement are shown in Fig. 61.

4.4.3 Dynamical properties

The Goldstone theorem predicts the existence of a gapless mode at long wavelength in the spectrum of a superfluid Fermi gas, which is associated with the broken gauge symmetry. In the BCS limit this is the Anderson-Bogoliubov phonon, which evolves into the Bogoliubov sound mode in the strong-coupling BEC limit. The velocity of the Goldstone phonon, the dynamic structure factor, and the amplitude of the velocity Bloch oscillations are possible observables to detect the onset of superfluidity and to distinguish the various regimes.

(i) Velocity of the Goldstone mode. In the BCS-BEC crossover there is a strongly hybridization between the Anderson-Bogoliubov phonon and the Bogoliubov phonon if the atoms are coupled to a Feshbach resonance. In such a case the gas is described by the Hamiltonian in Eq. (139). The dynamical properties of the gas can be studied by evaluating a set $\chi_{ij}(\mathbf{q}, \omega)$ of correlation functions ($i, j = 1, 2, 3$) [502]. Physically, χ_{11} and χ_{22} describe the amplitude and phase fluctuations of Cooper pairs, while χ_{33} describes density fluctuations in the gas of Fermi atoms and χ_{ij} with $i \neq j$ account for the couplings between the various fluctuations. The Bogoliubov density correlation functions are given by

$$\chi_{ij}^0(\mathbf{q}, \omega) = \frac{1}{\beta V} \sum_{\mathbf{k}, \omega_m} \text{Tr} \left[\tau_i \hat{G} \left(\mathbf{k} + \frac{\mathbf{q}}{2}, i\omega_m + i\omega \right) \times \tau_j \hat{G} \left(\mathbf{k} - \frac{\mathbf{q}}{2}, i\omega_m \right) \right].$$

$$(143)$$

Here τ_i are the Pauli matrices, ω_m are the Matsubara frequencies, and $\hat{G}(\mathbf{k}, i\omega)$ is a 2×2 matrix whose elements are the fermion thermal Green's functions $G_{\sigma\sigma'}(\mathbf{k}, i\omega)$ for the spin components at mean-field level. The effect of the interactions is then incorporated by a generalized RPA to evaluate $\chi_{ij}(\mathbf{q}, \omega)$ and hence the dynamical properties of the superfluid through the BCS-BEC crossover.

The correlation functions $\chi_{ij}(\mathbf{q}, \omega)$ exhibit the Goldstone mode as a pole having a phonon dispersion $\omega = cq$ in the long wavelength limit. For $T \to 0$ in the BCS regime ($\nu \gg \mu$), the region near the Fermi surface dominates and the sound velocity has the same value as first sound in a Fermi gas, namely $c = v_F/\sqrt{3}$. In the BEC regime where the non-resonant interaction g is strong, the sound velocity has the Bogoliubov form $c = (n_B|g_B|/M)^{1/2}$ where $n_B = N/2V$ is the density of composite bosons and $g_B = 4\pi\hbar^2 a_B/M$ is an effective coupling between bosons of mass $M = 2m$ with an s-wave scattering length $a_B = 2a$. The numerical results for the speed c of the Goldstone phonon in the crossover are shown in Fig. 62 and compared with the asymptotic expressions given above.

(ii) Dynamic structure factor. The density fluctuations spectrum of the atomic gas related to $\text{Im}\chi_{33}(\mathbf{q}, \omega)$ by the fluctuation-dissipation theorem and can be probed by scattering techniques as already discussed in Sec. 1.3.5. For the homogeneous system the evolution of the spectrum in the crossover has been studied by Ohashi and Griffin [502].

A local-density approximation can be used for a gas in harmonic confinement and the dynamic structure factor can be written as

$$S_{LDA}(\mathbf{q}, \omega) = \frac{-2\hbar}{1 - e^{-\beta\hbar\omega}} \int d^3 r \text{Im}\chi_{33}(\mathbf{q}, \omega; \mu(\mathbf{r}), \Delta(\mathbf{r})). \quad (144)$$

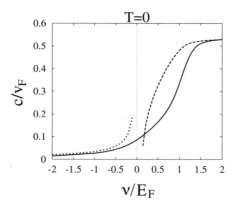

Fig. 62. Velocity c of the Goldstone phonon in the BCS-BEC crossover at $T = 0$ (solid line), compared with the expressions for the BCS limit (light-dashed line) and for the BEC limit (bold-dashed line). Redrawn from Ohashi and Griffin [502].

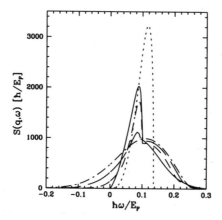

Fig. 63. Dynamic structure factor of a confined superfluid Fermi gas for $qa_{ho} = 4.15$ at temperatures increasing from $T = 0$ (solid line) to $k_BT = 0.042E_F$ (dot-dashed line). The dotted curve shows the the broadening of the Bogoliubov-Anderson phonon in the low-q and zero-temperature limit. Redrawn from Minguzzi $et\ al.$ [191].

In the BCS limit the spectrum of the trapped gas has been calculated by Minguzzi $et\ al.$ [191]. For this weak-coupling regime the numerical results in Fig. 63 show that the Anderson-Bogoliubov phonon is visible at low temperatures.

(iii) Bloch oscillations of a Fermi gas in an optical lattice. Fermions in a lattice can perform Bloch oscillations under the action of a constant force if the band is not completely filled. All fermions in a normal Fermi gas move incoherently and the oscillations are strongly damped, as experimentally ob-

served [503]. Cooper pairs in a superfluid should instead oscillate coherently and the damping could be negligible as for a BEC [89].

In a simulated experiment Rodriguez and Törmä [489] have driven the transition towards the superfluid state by varying the well depth in a lattice at constant temperature, as the critical temperature depends on this parameter (see Sec. 4.4.1). They have solved the Hamiltonian in Eq.(138) at mean-field level by using the Bogoliubov-de Gennes equations [504]. Their numerical results for the amplitude of the Bloch oscillations as a function of the well depth U_0 are shown in Fig. 58. The amplitude decreases with increasing U_0 as the fermions become localized and a sharp change in the amplitude is predicted to occur at the transition point.

4.4.4 High-T_c superfluids: the pseudogap phase

The concept of a pseudogap phase has been introduced in the context of high-T_c superconductors (see e.g. Ref. [505]) to characterize the intermediate-coupling regime beyond the model of Noziéres and Schmitt-Rink [183] (see Sec. 1.4.4). The same ideas are beginning to be exploited for cold atomic gases [506].

The pesudogap manifests itself below a certain temperature T_F^* as a dip in the density of states of the Fermi gas close to the chemical potential. Its appearance is linked to a breakdown of the picture of well-defined single-particle excitations near the Fermi surface. Local fermion-pair resonances emerge at a second characteristic temperature $T_B^* < T_F^*$. These two-fermion excitations condense at $T_c \ll T_B^*$ into a macroscopic quantum state. The opening of the pseudogap is thus driven by amplitude correlations connected with the dynamical formation of fluctuating pairs. Chiofalo et al. [507] have argued that, when about half of the total available fermions are involved in pair formation, these bosons substantially influence the physical properties of the normal state and could be linked to the minimum in the electrical resistivity which is observed [508] in undoped high-T_c cuprate superconductors.

While in the BCS limit the short-range amplitude correlations which are responsible for fermion pairing occur together with long-range phase coherence, phase correlations for $T_B^* < T < T_F^*$ are short-range and one can depict the system as a dynamically fluctuating mixture of fermion pairs and fermion monomers [509]. The role of fluctuations suggests that the nature of the pseudogap phase depends on dimensionality: while in 1D and 2D a well defined temperature range has been predicted [505], in 3D such a phase should be restricted to a very narrow temperature regime above T_c [510]. This emphasizes the importance of the quasi-2D nature of the cuprate materials, where the pseudogap above T_c has been observed by angle-resolved photoemission

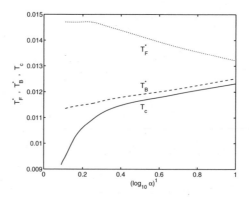

Fig. 64. Variation of T_F^*, T_B^* and T_c as functions of the anisotropy α in high-T_c cuprate superconductors. Redrawn from Devillard and Ranninger [509].

spectroscopy (see for example Ref. [511]).

The behaviour of T_F^*, T_B^* and T_c as functions of the anisotropy parameter $\alpha = m_\perp/m_\parallel$ characterizing the quasi-2D system, with $m_\perp$ and $m_\parallel$ being the orthogonal and logitudinal effective masses, has been calculated by Devillard and Ranninger [509] and is shown in Fig. 64. The system is described by a boson-fermion Hamiltonian which is the same as in Eq. (139) apart for the fact that the term involving the normal coupling g is omitted. In their model the fermionic and bosonic Green's functions $G_F(\mathbf{q}, \omega) = [\hbar\omega - \epsilon_\mathbf{q} - \Sigma_F(\mathbf{q}, \omega)]^{-1}$ and $G_B(\mathbf{q}, \varepsilon) = [\hbar\omega - \Delta + 2\mu - \Sigma_B(\mathbf{q}, \omega)]^{-1}$ are calculated self-consistently in a one-loop approximation for the self-energies $\Sigma_F(\mathbf{q}, \omega)$ and $\Sigma_B(\mathbf{q}, \omega)$. In the calculation T_F^* is inferred from the fermionic density of states, T_B^* is determined by the condition that the ratio between the imaginary and the real part of the boson self-energy vanishes, while T_c is determined from the Hugenholtz-Pines theorem by the condition $\mathrm{Re}\Sigma_B(0, 0) = \Delta - 2\mu$.

Direct calculations of the fermionic single-particle spectral function have been carried out by Perali et al. [486] at $T > T_c$ from weak to strong coupling. They have considered a 3D gas of fermions mutually interacting via an effective short-range potential characterized by the fermionic scattering length, without including the composite bosons in the Hamiltonian. The fermionic spectral function is calculated self-consistently using an expression for the self-energy at the one-loop level and displays two peaks which evolve with temperature. As in Ref. [509] two characteristic temperatures T_F^* and T_0^* can be identified: at $T = T_F^*$ the two peaks merge into one, while at $T = T_0^*$ their separation reaches its maximum width. Within their model the authors observe that T_F^* and T_0^* are well separated at strong coupling, while they both coincide with T_c in the weak-to-intermediate coupling regime.

The physics of pseudogap formation has been examined in a number of the-

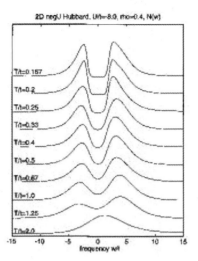

Fig. 65. Single-particle density of states $N(\omega)$ as a function of the energy ω (in units of the hopping energy t) for $U/t = -8.0$ (intermediate-to-strong coupling) and for a filling density $\rho = 0.4$. From Singer *et al.* [513].

oretical studies using the 2D Hubbard model with attractive interactions (see Eq. (138)). This is in general solved by the path-integral approach (see Sec. 3.1.2). An exhaustive discussion may be found in the review by Loktev *et al.* [512]. As an illustrative example we report in Fig. 65 QMC results for the pseudogap in such a model at various values of the temperature in the intermediate-to-strong coupling regime [513]. At high temperature only one peak is present in the density of states, and the pseudogap opens up as temperature is lowered.

Reference may also be made in closing to the use of Density Matrix Renormalization Group techniques to treat interacting quantum systems on a lattice [514,515].

5 Summary

Since the realization of the first Bose-Einstein condensates in 1995, the study of Bose and Fermi gases has expanded enormously to straddle fields of physics ranging from optics and quantum computing to superfluidity and quantum phase transitions. The rate at which knowledge has grown in these areas of quantum statistical mechanics is amazing. A few years before the time of writing (December 2003), the condensed state had already been fully characterized and the focus of interest was moving towards the study of vortices in trapped condensates, which had been newly generated by properly "stirring" the superfluid gas in the laboratory. At the time of writing the focus is on the newly discovered phenomena of superfluidity in trapped Fermi gases and Bose-Einstein condensation of tightly bound fermion pairs. One may safely anticipate for the immediate future a continued growth of interest in gases of strongly interacting fermions and in collective properties of vortex assemblies such as vortex phase diagrams in rotating condensates [317, 516] and vortex dynamics approaching the Quantum Hall regime [517, 518]. At the same time attention is being drawn to the emergence of a roton minimum in the dispersion relation of excitations in condensates with dipolar interactions [519, 520] and progress is being made in the realization and study of Bose-Einstein condensation in a gas of excitons [521]. Useful introductory references to further areas of main potential growth may be to a topical collection of articles on quantum computing [522], to studies of quantum turbulence and the role of vortex tangles [523-525], and to analogies between superfluidity and cosmological phenomena [526, 527].

The richness of the field is reflected by the wide array of computational tools which are being developed to address the great variety of physical situations. On the whole it seems fair to say that, given the very rapid development of the field, the emphasis has been more on "what to do" than on a systematization of "how to do it". In most cases, as we have seen, these tools are essentially a customized adaptation to the quantum-gas framework of techniques developed in other sectors of computational physics. Some other applications, and especially those involving the coupling of mean-field theory with classical and quantum kinetic theory, require a boost in sophistication. The possibility to describe the self-consistent evolution of a Bose condensate and an accompanying cloud of bosonic/fermionic thermal excitations depends on numerical schemes that may be capable of handling multiple and widely separated time scales.

Another easy prediction is the extension of present numerical methods in the direction of parallel computing. Given the actual capabilities of present-day parallel computers, boosts of two orders of magnitude in size and time-span can easily be gained for many quantum-gas applications with relatively minor programming efforts.

A more specific commentary can be made by separately summarizing the contents of the last three Chapters of this review. In Chapter 2 we have seen that the problem of computing the ground state of a BEC as described by the GPE is successfully handled by importing standard techniques for the solution of nonlinear Schrödinger equations. Given the wide array of techniques available for this purpose, one may ask which one works best for the problem at hand. A few general guidelines can be identified. A first major option arises in connection with numerical eigenvalue solvers for the time-independent GPE and evolution in imaginary time. Numerical eigensolvers are generally computationally intensive unless specific simplifications can be found, and dynamic integration of the time-dependent GPE is an appealing alternative. Here again a major option arises between explicit and implicit methods. The latter are generally recommended for steady-state problems, where no information is required on dynamical transients (excited states). If dynamics needs to be tracked, then explicit methods offer better accuracy at the cost of using small time steps, scaling with the square of the grid spacing as dictated by stability conditions. An additional bonus of explicit methods is their simplicity (no matrix algebra), and modern relaxation methods might help by offering stable operation with timesteps scaling only linearly with the grid spacing. For periodic geometries a very efficient solution is offered by time-splitting spectral methods, which combine good accuracy with efficiency and ease of use. Further benefits from more advanced methodologies can be envisaged in the areas of (i) adaptive grids that are widely used in computational fluid dynamics to resolve flows with sharp features and may be usefully extended to quantum fluids [528, 529]; (ii) multigrid methods, in which the solution is evolved on a set of hierarchical grids allowing selective removal of errors of different wavelengths by means of grids of different fitness; and (iii) finite-volume and finite-element methods, allowing more complicated geometries to be handled as in problems of nuclear fusion [232].

The GPE has also been useful to deal with a number of physical situations in which a dilute Bose-Einstein condensate is accompanied by a cloud of bosonic or fermionic atoms. These cases require the numerical solution of the GPE in conjunction with the solution of the VLE. The many numerical methods that are available in modern physics do not as a rule straddle across classical and quantum aspects in the way studies of ultra-cold quantum gas do. The methods described in Chapter 3 to deal with Bose gases at finite temperature range from attempts to stay close to the familiar and efficient GPE-based framework to fairly sophisticated functional techniques for stochastic versions of the GPE. A quite common trait of these methods is some form of static or dynamic Monte Carlo sampling in phase space, with those special requirements of enhanced efficiency that arise from dealing with different time scales for quantum and semi-classical degrees of freedom and from the quenching of collisions by quantum effects in ultra-cold Fermi gases. The latter effect has been amply illustrated by our discussion of colliding clouds of fermions given in Chapter 4 on the basis of two coupled VLE's.

Functional methods have been developed to deal with thermal and quantum fluctuations in a Bose gas on an equal footing, as discussed in Chapter 3. These methods shine for their formal elegance and may provide unique clues into the nature of both classical and quantum field theories. Their viability as practical numerical tools almost invariably hinges, however, on the availability of a stable zero-point (Gaussian) configuration to expand about. Such stable configurations are indeed available for Bose-Einstein condensates, so that sufficiently weak fluctuations can be successfully handled by functional integration. Extending this approach to strong-fluctuating regimes is a formidable challenge, but any progress along this line could provide significant progress for many unsolved problems in classical and quantum nonlinear field theories and not only for BEC physics.

At variance from the GPE framework, numerical analysis is of little help in dealing with strongly correlated quantum gases in situations such as are met in studying many-body effects in Bose-Einstein condensates at zero or finite temperature and in strongly correlated Fermi gases. Much has been borrowed from existing methods of quantum statistical physics both in terms of models and of computational methods, the prime tools for numerical studies in these areas being again the Diffusion Monte Carlo and Path-Integral Monte Carlo methods. Continued progress in these areas will surely benefit the study of quantum gases in years to come.

Acknowledgements

We wish to thank on this occasion all the numerous friends with whom we have over the years had illuminating discussions on the topics covered in this review. We are especially indebted to Professor B. V. Svistunov for a critical reading of the manuscript and for his suggestions. We also acknowledge the support given to us through specific Advanced Research Programs by the Italian Ministry for University and Research (MIUR) and by the Istituto Nazionale di Fisica della Materia (INFM).

References

[1] A. S. Parkin, D. F. Walls, The physics of trapped dilute-gas Bose-Einstein condensates, Phys. Rep. 303 (1998) 1.

[2] F. Dalfovo, S. Giorgini, L. Pitaevskii, S. Stringari, Theory of Bose-Einstein condensation in trapped gases, Rev. Mod. Phys. 71 (1999) 463.

[3] A. J. Leggett, Bose-Einstein condensation in alkali gases: Some fundamental concepts, Rev. Mod. Phys. 73 (2001) 307.

[4] C. J. Pethick, H. Smith, Bose-Einstein Condensation in Atomic Gases, Cambridge University Press, Cambridge, 2002.

[5] L. P. Pitaevskii, S. Stringari, Bose-Einstein Condensation, Clarendon, Oxford, 2003.

[6] M. Inguscio, S. Stringari, C. Wieman (Eds.), Bose-Einstein condensation in atomic gases, Proc. Int. School "Enrico Fermi", IOS Press, Amsterdam, 1999.

[7] S. Martellucci, A. N. Chester, A. Aspect, M. Inguscio (Eds.), Bose-Einstein condensates and atom lasers – Proc. Int. School of Quantum Electronics, Kluwer, New York, 2000.

[8] A. Aspect, J. Dalibard (Eds.), Comptes Rendus de l'Académie des Sciences–Dossier: Bose-Einstein condensates and atom lasers, Paris, 2001.

[9] A. Aspect, J. Dalibard (Eds.), Comptes Rendus de l'Académie des Sciences–Dossier: Atom optics and interferometry, Paris, 2001.

[10] R. Kaiser, C. Westbrook, F. David (Eds.), Coherent atomic matter waves – Les Houches Ècole de Physique – Ècole d'été de Physique Theorique, Session LXXII, Springer, Berlin, 2001.

[11] H. R. Sadeghpour, J. L. Bohn, M. J. Cavagnero, B. D. Esry, I. I. Fabrikant, J. H. Macek, A. R. P. Rau, Collisions near threshold in atomic and molecular physics, J. Phys. B 33 (2000) R93.

[12] A. A. Abrikosov, L. P. Gorkov, I. E. Dzyaloshinski, Methods of Quantum Field Theory in Statistical Physics, Dover, New York, 1963.

[13] V. K. Wong, H. Gould, Long-wavelength excitations in a Bose gas at zero temperature, Ann. Phys. (NY) 83 (1974) 253.

[14] D. S. Petrov, M. Holzmann, G. V. Shlyapnikov, Bose-Einstein condensation in quasi-2D trapped gases, Phys. Rev. Lett. 84 (2000) 2551.

[15] M. Olshanii, Atomic scattering in the presence of an external confinement and a gas of impenetrable bosons, Phys. Rev. Lett. 81 (1998) 938.

[16] N. Bogoliubov, On the theory of superfluidity, J. Phys. USSR 11 (1947) 23.

[17] J. Gavoret, P. Nozières, Structure of the perturbation expansion for the Bose liquid at zero temperature, Ann. Phys. (N.Y.) 28 (1964) 349.

[18] Y. Castin, R. Dum, Low-temperature Bose-Einstein condensates in time-dependent traps: Beyond the U(1) symmetry-breaking approach, Phys. Rev. A 57 (1998) 3008.

[19] N. M. Hugenholtz, D. Pines, Ground-state energy and excitation spectrum of a system of interacting bosons, Phys. Rev. 116 (1959) 489.

[20] P. C. Hohenberg, P. C. Martin, Microscopic theory of superfluid Helium, Ann. Phys. (N.Y.) 34 (1965) 291.

[21] O. Penrose, L. Onsager, Bose-Einstein condensation and liquid Helium, Phys. Rev. 104 (1956) 576.

[22] J. F. Allen, A. D. Misener, Flow of liquid Helium II, Nature 141 (1938) 75.

[23] V. P. Peshkov, Second sound in He II, J. Phys. Moscow 8 (1944) 381.

[24] G. B. Hess, W. M. Fairbank, Measurements of angular momentum in superfluid Helium, Phys. Rev. Lett. 19 (1967) 216.

[25] L. Tisza, Transport phenomena in Helium II, Nature 141 (1938) 913.

[26] K. M. Khalatnikov, An Introduction to the Theory of Superfluidity, Benjamin, New York, 1965.

[27] E. L. Andronikashvili, Y. G. Malamadze, Quantization of macroscopic motions and hydrodynamics of rotating Helium II, Rev. Mod. Phys. 38 (1966) 567.

[28] K. R. Atkins, Liquid Helium, Cambridge University Press, London, 1959.

[29] H. R. Glyde, R. T. Azuah, W. G. Stirling, Condensate, momentum distribution, and final-state effects in liquid ^{4}He, Phys. Rev. B 62 (2000) 14337.

[30] W. N. Snow, Y. Wang, P. E. Sokol, Density and temperature dependence of the condensate fraction in liquid ^{4}He, Europhys. Lett. 19 (1992) 403.

[31] S. Moroni, G. Senatore, S. Fantoni, Momentum distribution of liquid Helium, Phys. Rev. B 55 (1997) 1040.

[32] P. A. Whitlock, R. Panoff, Accurate momentum distributions from computations on ^{3}He and ^{4}He, Can. J. Phys. 65 (1987) 1409.

[33] E. Manousakis, V. R. Pandharipande, Q. N. Usmani, Condensate fraction and momentum distribution in the ground state of liquid ^{4}He, Phys. Rev. B 31 (1985) 7022.

[34] D. M. Ceperley, E. L. Pollock, Path-integral computation of the low-temperature properties of liquid ^{4}He, Phys. Rev. Lett. 56 (1986) 351.

[35] S. Grebenev, J. P. Toennies, A. F. Vilesov, Superfluidity within a small Helium-4 cluster: The microscopic Andronikashvili experiment, Science 279 (1998) 2083.

[36] C. Callegari, A. Conjusteau, I. Reinhard, K. K. Lehmann, G. Scoles, Superfluid hydrodynamic model for the enhanced moments of inertia of molecules in liquid ^{4}He, Phys. Rev. Lett. 83 (1999) 5058.

[37] E. W. Draeger, D. M. Ceperley, Superfluidity in a doped Helium droplet, Phys. Rev. Lett. 89 (2002) 015301.

[38] B. D. Josephson, Relation between the superfluid density and order parameter for superfluid He near T_c, Phys. Lett. 21 (1966) 608.

[39] V. N. Popov, Functional Integrals in Quantum Field Theory and Statistical Physics, Reidel, Dordrecht, 1983.

[40] G. Baym, Microscopic theory of superfluid Helium, in: R. C. Clark, G. H. Derrick (Eds.), Mathematical Methods in Solid State and Superfluid Theory, Oliver and Boyd, Edinburgh, 1969, p. 121.

[41] A. J. Leggett, Superfluidity, Rev. Mod. Phys. 71 (1999) S318.

[42] P. C. Hohenberg, Existence of long-range order in one and two dimensions, Phys. Rev. 158 (1966) 383.

[43] W. R. Magro, D. M. Ceperley, Ground-state properties of the two-dimensional Bose Coulomb liquid, Phys. Rev. Lett. 73 (1994) 826.

[44] L. Pitaevskii, S. Stringari, Uncertainty principle, quantum fluctuations and broken symmetries, J. Low Temp. Phys. 85 (1991) 377.

[45] D. S. Fisher, P. C. Hohenberg, Dilute Bose gas in two dimensions, Phys. Rev. B 37 (1988) 4936.

[46] J. T. M. Walraven, E. R. Eliel, I. F. Silvera, Experimental study of spin aligned atomic Hydrogen condensed on surfaces, Phys. Lett. A 66 (1978) 247.

[47] T. Hänsch, A. Schawlow, Cooling of gases by laser radiation, Opt. Commun. 13 (1975) 68.

[48] J. Dalibard, C. Cohen-Tannoudji, Laser cooling below the Doppler limit by polarization gradients: Simple theoretical models, J. Opt. Soc. Am. B 6 (1989) 2023.

[49] A. Aspect, E. Arimondo, R. Kaiser, N. Vansteenkiste, C. Cohen-Tannoudji, Laser cooling below the one-photon recoil energy by velocity-selective coherent population trapping, Phys. Rev. Lett. 61 (1988) 826.

[50] M. A. Olshanii, V. G. Minogin, Three-dimensional velocity-selective coherent population trapping of (3+1)-level atoms, Quantum Opt. 3 (1991) 317.

[51] J. Lawall, S. Kulin, B. Saubaméa, N. Bigelow, M. Leduc, C. Cohen-Tannoudji, Three-dimensional laser cooling of Helium beyond the single-photon recoil limit, Phys. Rev. Lett. 75 (1995) 4194.

[52] W. Petrich, M. H. Anderson, J. R. Ensher, E. A. Cornell, Stable, tightly confining magnetic trap for evaporative cooling of neutral atoms, Phys. Rev. Lett. 74 (1995) 3352.

[53] M. H. Anderson, J. R. Ensher, M. R. Matthews, C. E. Wieman, E. A. Cornell, Observation of Bose-Einstein condensation in a dilute atomic vapor, Science 269 (1995) 198.

[54] E. A. Cornell, C. E. Wieman, Nobel Lecture: Bose-Einstein condensation in a dilute gas, the first 70 years and some recent experiments, Rev. Mod. Phys. 74 (2002) 875.

[55] W. Ketterle, Nobel Lecture: When atoms behave as waves: Bose-Einstein condensation and the atom laser, Rev. Mod. Phys. 74 (2002) 1131.

[56] K. B. Davis, M. O. Mewes, M. R. Andrews, N. J. van Druten, D. S. Durfee, D. M. Kurn, W. Ketterle, Bose-Einstein condensation in a gas of sodium atoms, Phys. Rev. Lett. 75 (1995) 3969.

[57] C. C. Bradley, C. A. Sackett, J. J. Tollett, R. G. Hulet, Evidence of Bose-Einstein condensation in an atomic gas with attractive interactions, Phys. Rev. Lett. 75 (1995) 1687, ibid. 79 (1997) 1170.

[58] D. G. Fried, T. C. Killian, L. Willmann, D. Landhuis, S. C. Moss, D. Kleppner, T. J. Greytak, Bose-Einstein condensation of atomic Hydrogen, Phys. Rev. Lett. 81 (1998) 3811.

[59] S. L. Cornish, N. R. Claussen, J. L. Roberts, E. A. Cornell, C. E. Wieman, Stable ^{85}Rb Bose-Einstein condensates with widely tunable interactions, Phys. Rev. Lett. 85 (2000) 1795.

[60] A. Robert, O. Sirjean, A. Browaeys, J. Poupard, S. Nowak, D. Boiron, C. I. Westbrook, A. Aspect, A Bose-Einstein condensate of metastable atoms, Science 292 (2001) 461.

[61] F. P. D. Santos, J. Léonard, J. Wang, C. J. Barrelet, F. Perales, E. Rasel, C. S. Unnikrishnan, M. Leduc, C. Cohen-Tannoudji, Bose-Einstein condensation of metastable Helium, Phys. Rev. Lett. 86 (2001) 3459.

[62] G. Modugno, G. Ferrari, G. Roati, R. J. Brecha, A. Simoni, M. Inguscio, Bose-Einstein condensation of Potassium atoms by sympathetic cooling, Science 294 (2001) 1320.

[63] T. Weber, J. Herbig, M. Mark, H.-C. Nägerl, R. Grimm, Bose-Einstein condensation of Cesium, Science 299 (2003) 232.

[64] Y. Takasu, K. Maki, K. Komori, T. Takano, K. Honda, M. Kumakura, T. Yabuzaki, Y. Takahashi, Spin-singlet Bose-Einstein condensation of two-electron atoms, Phys. Rev. Lett. 91 (2003) 040404.

[65] J. R. Ensher, D. S. Jin, M. R. Matthews, C. E. Wieman, E. A. Cornell, Bose-Einstein condensation in a dilute gas: Measurement of energy and ground-state occupation, Phys. Rev. Lett. 77 (1996) 4984.

[66] S. Giorgini, L. P. Pitaevskii, S. Stringari, Scaling and thermodynamics of a trapped Bose-condensed gas, Phys. Rev. Lett. 78 (1997) 3987.

[67] E. A. Cornell, J. R. Ensher, C. E. Wieman, Experiments in dilute atomic Bose-Einstein condensation, in: M. Inguscio, S. Stringari, C. E. Wieman (Eds.), Proc. Int. School "Enrico Fermi", IOS Press, Amsterdam, 1999, p. 15.

[68] S. Giorgini, L. P. Pitaevskii, S. Stringari, Condensate fraction and critical temperature of a trapped interacting Bose gas, Phys. Rev. A 54 (1996) R4633.

[69] F. Gerbier, J. H. Thywissen, S. Richard, M. Hugbart, P. Bouyer, A. Aspect, Critical temperature of a trapped weakly interacting Bose gas Cond-mat/0307188.

[70] W. Kohn, Cyclotron resonance and de Haas-van Alphen oscillations of an interacting electron gas, Phys. Rev. 123 (1961) 1242.

[71] J. F. Dobson, Harmonic-potential theorem: Implications for approximate many-body theories, Phys. Rev. Lett. 73 (1994) 2244.

[72] D. S. Jin, J. R. Ensher, M. R. Matthews, C. E. Wieman, E. A. Cornell, Collective excitations of a Bose-Einstein condensate in a dilute gas, Phys. Rev. Lett. 77 (1996) 420.

[73] M. O. Mewes, M. R. Andrews, N. J. van Druten, D. M. Kurn, D. S. Durfee, C. G. Townsend, W. Ketterle, Collective excitations of a Bose-Einstein condensate in a magnetic trap, Phys. Rev. Lett. 77 (1996) 988.

[74] R. Onofrio, D. S. Durfee, C. Raman, M. Köhl, C. E. Kuklewicz, W. Ketterle, Surface excitations of a Bose-Einstein condensate, Phys. Rev. Lett. 84 (2000) 810.

[75] A. Griffin, W.-C. Wu, S. Stringari, Hydrodynamic modes in a trapped Bose gas above the Bose-Einstein transition, Phys. Rev. Lett. 78 (1997) 1838.

[76] D. S. Jin, M. R. Matthews, J. R. Ensher, C. E. Wieman, E. A. Cornell, Temperature-dependent damping and frequency shifts in collective excitations of a dilute Bose-Einstein condensate, Phys. Rev. Lett. 78 (1997) 764.

[77] D. M. Stamper-Kurn, H. J. Miesner, S. Inouye, M. R. Andrews, W. Ketterle, Collisionless and hydrodynamic excitations of a Bose-Einstein condensate, Phys. Rev. Lett. 81 (1998) 500.

[78] M. J. Bijlsma, H. T. C. Stoof, Collisionless modes of a trapped Bose gas, Phys. Rev. A 60 (1999) 3973.

[79] J. Reidl, A. Csordás, R. Graham, P. Szépfalusy, Shifts and widths of collective excitations in trapped Bose gases determined by the dielectric formalism, Phys. Rev. A 61 (2000) 043606.

[80] B. Jackson, E. Zaremba, Quadrupole collective modes in trapped finite-temperature Bose-Einstein condensates, Phys. Rev. Lett. 88 (2002) 180402.

[81] S. A. Morgan, M. Rusch, D. A. W. Hutchinson, K. Burnett, Quantitative test of thermal field theory for Bose-Einstein condensates Cond-mat/0305535.

[82] M. R. Andrews, D. M. Kurn, H. J. Miesner, D. S. Durfee, C. G. Townsend, S. Inouye, W. Ketterle, Propagation of sound in a Bose-Einstein condensate, Phys. Rev. Lett. 79 (1997) 553, *ibid.* 80 (1998) 2967.

[83] G. Hechenblaikner, O. M. Maragò, E. Hodby, J. Arlt, S. Hopkins, C. J. Foot, Observation of harmonic generation and nonlinear coupling in the collective dynamics of a Bose-Einstein condensate, Phys. Rev. Lett. 85 (2000) 692.

[84] S. Burger, K. Bongs, S. Dettmer, W. Ertmer, K. Sengstock, A. Sanpera, G. V. Shlyapnikov, M. Lewenstein, Dark solitons in Bose-Einstein condensates, Phys. Rev. Lett. 83 (1999) 5198.

[85] J. Denschlag, J. E. Simsarian, D. L. Feder, C. W. Clark, L. A. Collins, J. Cubizolles, L. Deng, E. W. Hagley, K. Helmerson, W. P. Reinhardt, S. L. Rolston, B. I. Schneider, W. D. Phillips, Generating solitons by phase engineering of a Bose-Einstein condensate, Science 287 (2000) 97.

[86] B. P. Anderson, P. C. Haljan, C. A. Regal, D. L. Feder, L. A. Collins, C. W. Clark, E. A. Cornell, Watching dark solitons decay into vortex rings in a Bose-Einstein condensate, Phys. Rev. Lett. 86 (2001) 2926.

[87] Z. Dutton, M. Budde, C. Slowe, L. Hau, Observation of quantum shock waves created with ultra- compressed slow light pulses in a Bose-Einstein condensate, Science 293 (2001) 663.

[88] M. O. Mewes, M. R. Andrews, D. M. Kurn, D. S. Durfee, C. G. Townsend, W. Ketterle, Output coupler for Bose-Einstein condensed atoms, Phys. Rev. Lett. 78 (1997) 582.

[89] B. P. Anderson, M. A. Kasevich, Macroscopic quantum interference from atomic tunnel arrays, Science 282 (1998) 1686.

[90] I. Bloch, T. W. Hänsch, T. Esslinger, Atom laser with a cw output coupler, Phys. Rev. Lett. 82 (1999) 3008.

[91] E. W. Hagley, L. Deng, M. Kozuma, J. Wen, K. Helmerson, S. L. Rolston, W. D. Phillips, A well-collimated quasi-continuous atom laser, Science 283 (1999) 1706.

[92] G. Cennini, G. Ritt, C. Geckeler, M. Weitz, All-optical realization of an atom laser, Phys. Rev. Lett. 91 (2003) 240408.

[93] Y. Kagan, B. V. Svistunov, Kinetics of the onset of long-range order during Bose condensation in an interacting gas, Sov. Phys. JETP 78 (1994) 187.

[94] Y. Kagan, Kinetics of Bose-Einstein condensate formation in an interacting Bose gas, in: A. Griffin, D. W. Snoke, S. Stringari (Eds.), Bose-Einstein Condensation, Cambridge University Press, Cambridge, 1995, p. 202.

[95] H. T. C. Stoof, Initial stages of Bose-Einstein condensation, Phys. Rev. Lett. 78 (1997) 768.

[96] C. W. Gardiner, P. Zoller, R. J. Ballagh, M. J. Davis, Kinetics of Bose-Einstein condensation in a trap, Phys. Rev. Lett. 79 (1997) 1793.

[97] H. J. Miesner, D. M. Stamper-Kurn, M. R. Andrews, D. S. Durfee, S. Inouye, W. Ketterle, Bosonic stimulation in the formation of a Bose-Einstein condensate, Science 279 (1998) 1005.

[98] J. Javanainen, S. M. Yoo, Quantum phase of a Bose-Einstein condensate with an arbitrary number of atoms, Phys. Rev. Lett. 76 (1996) 161.

[99] K. Mølmer, Phase collapse and excitations in Bose-Einstein condensates, Phys. Rev. A 58 (1998) 566.

[100] M. R. Andrews, C. G. Townsend, H. J. Miesner, D. S. Durfee, D. M. Kurn, W. Ketterle, Observation of interference between two Bose-Einstein condensates, Science 275 (1997) 637.

[101] W. Ketterle, D. S. Durfee, D. M. Stamper-Kurn, Making, probing and understanding Bose-Einstein condensates, in: M. Inguscio, S. Stringari, C. E. Wieman (Eds.), Proc. Int. School "Enrico Fermi", IOS Press, Amsterdam, 1999, p. 67.

[102] Y. Castin, Bose-Einstein condensates in atomic gases, in: R. Kaiser, C. Westbrook, F. David (Eds.), Coherent atomic matter waves, Proceeding of Les Houches NATO Advanced Institute, Springer, Berlin, 2001, p. 5.

[103] A. Röhrl, M. Naraschewski, A. Schenzle, H. Wallis, Transition from phase locking to the interference of independent Bose condensates: Theory versus experiment, Phys. Rev. Lett. 78 (1997) 4143.

[104] M. Kozuma, L. Deng, E. W. Hagley, J. Wen, R. Lutwak, K. Helmerson, S. L. Rolston, W. D. Phillips, Coherent splitting of Bose-Einstein condensed atoms with optically induced Bragg diffraction, Phys. Rev. Lett. 82 (1999) 871.

[105] S. Inouye, T. Pfau, S. Gupta, A. P. Chikkatur, A. Görlitz, D. E. Pritchard, W. Ketterle, Phase-coherent amplification of atomic matter waves, Nature 402 (1999) 641.

[106] M. O. Mewes, M. R. Andrews, N. J. van Druten, D. M. Kurn, D. S. Durfee, W. Ketterle, Bose-Einstein condensation in a tightly confining dc magnetic trap, Phys. Rev. Lett. 77 (1996) 416.

[107] M. J. Holland, D. S. Jin, M. L. Chiofalo, J. Cooper, Emergence of interaction effects in Bose-Einstein condensation, Phys. Rev. Lett. 78 (1997) 3801.

[108] B. D. Busch, C. Liu, Z. Dutton, C. H. Behroozi, L. V. Hau, Observation of interaction dynamics in finite-temperature Bose condensed atom clouds, Europhys. Lett. 51 (2000) 485.

[109] E. A. Burt, R. W. Ghrist, C. J. Myatt, M. J. Holland, E. A. Cornell, C. E. Wieman, Coherence, correlations, and collisions: What one learns about Bose-Einstein condensates from their decay, Phys. Rev. Lett. 79 (1997) 337.

[110] Y. Kagan, B. V. Svistunov, G. V. Shlyapnikov, Effect of Bose condensation on inelastic processes in gases, JEPT Lett. 42 (1985) 209.

[111] A. I. Safonov, S. A. Vasilyev, I. S. Yasnikov, I. I. Lukashevich, S. Jaakkola, Observation of quasicondensate in two-dimensional atomic Hydrogen, Phys. Rev. Lett. 81 (1998) 4545.

[112] Y. Kagan, B. V. Svistunov, G. V. Shlyapnikov, Influence on inelastic processes of the phase transition in a weakly collisional two-dimensional Bose gas, Sov. Phys. JEPT 66 (1987) 314.

[113] Y. Kagan, V. A. Kashurnikov, A. V. Krasavin, N. V. Prokof'ev, B. V. Svistunov, Quasicondensation in a two-dimensional interacting Bose gas, Phys. Rev. A 61 (2000) 43608.

[114] O. M. Maragó, S. A. Hopkins, J. Arlt, E. Hodby, G. Hechenblaikner, C. J. Foot, Observation of the scissors mode and evidence for superfluidity of a trapped Bose-Einstein condensed gas, Phys. Rev. Lett. 84 (2000) 2056.

[115] D. Guéry-Odelin, S. Stringari, Scissors mode and superfluidity of a trapped Bose-Einstein condensed gas, Phys. Rev. Lett. 83 (1999) 4452.

[116] K. W. Madison, F. Chevy, W. Wohlleben, J. Dalibard, Vortex formation in a stirred Bose-Einstein condensate, Phys. Rev. Lett. 84 (2000) 806.

[117] M. R. Matthews, B. P. Anderson, P. C. Haljan, D. S. Hall, C. E. Wieman, E. A. Cornell, Vortices in a Bose-Einstein condensate, Phys. Rev. Lett. 83 (1999) 2498.

[118] A. E. Leanhardt, A. Görlitz, A. P. Chikkatur, D. Kielpinski, Y. Shin, D. E. Pritchard, W. Ketterle, Imprinting vortices in a Bose-Einstein condensate using topological phases, Phys. Rev. Lett. 89 (2002) 190403.

[119] F. Chevy, K. W. Madison, V. Bretin, J. Dalibard, Formation of quantized vortices in a gaseous Bose-Einstein condensate, in: S. Atutov, C. Calabrese, L. Moi (Eds.), Proc. Particles and Fundamental Physics Workshop (Les Houches 2001), Kluwer, Dordrecht, 2002.

[120] J. R. Abo-Shaeer, C. Raman, J. M. Vogels, W. Ketterle, Observation of vortex lattices in Bose-Einstein condensate, Science 292 (2001) 476.

[121] A. L. Fetter, A. A. Svidzinsky, Vortices in a trapped dilute Bose-Einstein condensate, J. Phys.: Condens. Matter 13 (2001) R135.

[122] F. Chevy, K. W. Madison, J. Dalibard, Measurement of the angular momentum of a rotating Bose-Einstein condensate, Phys. Rev. Lett. 85 (2000) 2223.

[123] S. Inouye, S. Gupta, T. Rosenband, A. P. Chikkatur, A. Görlitz, T. L. Gustavson, A. E. Leanhardt, D. E. Pritchard, W. Ketterle, Observation of vortex phase singularities in Bose-Einstein condensates, Phys. Rev. Lett. 87 (2001) 080402.

[124] R. Onofrio, C. Raman, J. M. Vogels, J. R. Abo-Shaeer, A. P. Chikkatur, W. Ketterle, Observation of superfluid flow in a Bose-Einstein condensed gas, Phys. Rev. Lett. 85 (2000) 2228.

[125] A. P. Chikkatur, A. Görlitz, D. M. Stamper-Kurn, S. Inouye, S. Gupta, W. Ketterle, Suppression and enhancement of impurity scattering in a Bose-Einstein condensate, Phys. Rev. Lett. 85 (2000) 483.

[126] S. Burger, F. S. Cataliotti, C. Fort, F. Minardi, M. Inguscio, M. L. Chiofalo, M. P. Tosi, Superfluid and dissipative dynamics of a Bose-Einstein condensate in a periodic optical potential, Phys. Rev. Lett. 86 (2001) 4447.

[127] F. S. Cataliotti, L.Fallani, F. Ferlaino, C. Fort, P. Maddaloni, M.Inguscio, A. Smerzi, A. Trombettoni, P. G. Kevrekidis, A. R. Bishop, A novel mechanism for superfluidity breakdown in weakly coupled Bose-Einstein condensates Cond-mat/0207139.

[128] M. B. Dahan, E. Peik, J. Reichel, Y. Castin, C. Salomon, Bloch oscillations of atoms in an optical potential, Phys. Rev. Lett. 76 (1996) 4508.

[129] E. Peik, M. B. Dahan, I. Bouchoule, Y. Castin, C. Salomon, Bloch oscillations of atoms, adiabatic rapid passage, and monokinetic atomic beams, Phys. Rev. A 55 (1997) 2989.

[130] M. Greiner, I. Bloch, O. Mandel, T. W. Hänsch, T. Esslinger, Exploring phase coherence in a 2D lattice of Bose-Einstein condensates, Phys. Rev. Lett. 87 (2001) 160405.

[131] K. Berg-Sørensen, K. Mølmer, Bose-Einstein condensates in spatially periodic potentials, Phys. Rev. A 58 (1998) 1480.

[132] M. L. Chiofalo, M. Polini, M. P. Tosi, Collective excitations of a periodic Bose condensate in the Wannier representation, Eur. Phys. J. D 11 (2000) 371.

[133] M. L. Chiofalo, S. Succi, M. P. Tosi, Probing the energy bands of a Bose-Einstein condensate in a optical lattice, Phys. Rev.A 63 (2001) 063613.

[134] M. Kraemer, C. Menotti, L. P. Pitaevskii, S. Stringari, Bose-Einstein condensates in 1D optical lattices: Compressibility, Bloch bands and elementary excitations Cond-mat/0305300.

[135] B. Wu, Q. Niu, Nonlinear Landau-Zener tunneling, Phys. Rev. A 61 (2000) 023402.

[136] D. Diakonov, L. M. Jensen, C. J. Pethick, H. Smith, Loop structure of the lowest Bloch band for a Bose-Einstein condensate, Phys. Rev. A 66 (2002) 013604.

[137] M. Machholm, C. J. Pethick, H. Smith, Band structure, elementary excitations, and stability of a Bose-Einstein condensate in a periodic potential, Phys. Rev. A 67 (2003) 053613.

[138] W. Kohn, Analytic properties of Bloch waves and Wannier functions, Phys. Rev. 115 (1959) 809.

[139] J. Stenger, S. Inouye, A. P. Chikkatur, D. M. Stamper-Kurn, D. E. Pritchard, W. Ketterle, Bragg spectroscopy of a Bose-Einstein condensate, Phys. Rev. Lett. 82 (1999) 4569.

[140] J. Steinhauer, R. Ozeri, N. Katz, N. Davidson, Excitation spectrum of a Bose-Einstein condensate, Phys. Rev. Lett. 88 (2002) 120407.

[141] J. Steinhauer, N. Katz, R. Ozeri, N. Davidson, C. Tozzo, F. Dalfovo, Bragg spectroscopy of the multibranch Bogoliubov spectrum of elongated Bose-Einstein condensates, Phys. Rev. Lett. 90 (2003) 060404.

[142] S. Richard, F. Gerbier, J. H. Thywissen, P. B. M. Hugbart, A. Aspect, Momentum spectroscopy of 1D phase fluctuations in Bose-Einstein condensates, Phys. Rev. Lett. 91 (2003) 010405.

[143] E. W. Hagley, L. Deng, M. Kozuma, M. Trippenbach, Y. B. Band, M. Edwards, M. R. Doery, P. S. Julienne, K. Helmerson, S. L. Rolston, W. D. Phillips, Measurement of the coherence of a Bose-Einstein condensate, Phys. Rev. Lett. 83 (1999) 3112.

[144] P. Vignolo, Z. Akdeniz, M. P. Tosi, Transmittivity of a Bose-Einstein condensate on a lattice: interference from period doubling and the effect of disorder, J. Phys. B 36 (2003) 4535.

[145] M. L. Chiofalo, M. P. Tosi, Josephson-type oscillations of a driven Bose-Einstein condensate in an optical lattice, Europhys. Lett. 56 (2001) 326.

[146] F. S. Cataliotti, S. Burger, C. Fort, P. Maddaloni, F. Minardi, A. Trombettoni, A. Smerzi, M. Inguscio, Josephson junction arrays with Bose-Einstein condensates, Science 293 (2001) 843.

[147] C. Orzel, A. K. Tuchman, M. L. Fenselau, M. Yasuda, M. A. Kasevich, Squeezed states in a Bose-Einstein condensate, Science 291 (2001) 2386.

[148] M. P. A. Fisher, P. W. Weichmann, G. Grinstein, D. S. Fisher, Boson localization and the superfluid-insulator transition, Phys. Rev. B 40 (1989) 546.

[149] M. Greiner, O. Mandel, T. Esslinger, T. W. Hänsch, I. Bloch, Quantum phase transition from a superfluid to a Mott insulator in a gas of ultracold atoms, Nature 415 (2002) 39.

[150] D. Jaksch, C. Bruder, J. I. Cirac, P. Zoller, Cold bosonic atoms in optical lattices, Phys. Rev. Lett. 81 (1998) 3108.

[151] D. Jaksch, H.-J. Briegel, J. I. Cirac, C. W. Gardiner, P. Zoller, Entanglement of atoms via cold controlled collisions, Phys. Rev. Lett. 82 (1999) 1975.

[152] S. Peil, J. V. Porto, B. L. Tolra, J. M. Obrecht, B. E. King, M. Subbotin, S. L. Rolston, W. D. Phillips, Patterned loading of a Bose-Einstein condensate into an optical lattice, Phys. Rev. A 67 (2003) 051603.

[153] O. Mandel, M. Greiner, A. Widera, T. Rom, T. W. Hänsch, I. Bloch, Controlled collisions for multi-particle entanglement of optically trapped atoms, Nature 425 (2003) 937.

[154] E. Demler, F. Zhou, Spinor bosonic atoms in optical lattices: Symmetry breaking and fractionalization, Phys. Rev. Lett. 88 (2002) 163001.

[155] B. Damski, L. Santos, E. Tiemann, M. Lewenstein, S. Kotochigova, P. Julienne, P. Zoller, Creation of a dipolar superfluid in optical lattices, Phys. Rev. Lett. 90 (2003) 110401.

[156] B. DeMarco, D. S. Jin, Onset of Fermi degeneracy in a trapped atomic gas, Science 285 (1999) 1703.

[157] A. G. Truscott, K. E. Strecker, W. I. McAlexander, G. B. Partridge, R. G. Hulet, Observation of Fermi pressure in a gas of trapped atoms, Science 291 (2001) 2570.

[158] E. Timmermans, R. Côté, Superfluidity in sympathetic cooling with atomic Bose-Einstein condensates, Phys. Rev. Lett. 80 (1998) 3419.

[159] H. T. C. Stoof, M. Houbiers, C. A. Sackett, R. G. Hulet, Superfluidity of spin-polarized ^{6}Li, Phys. Rev. Lett. 76 (1996) 10.

[160] K. E. Strecker, G. B. Partridge, R. G. Hulet, Conversion of an atomic Fermi gas to a long-lived molecular Bose gas, Phys. Rev. Lett. 91 (2003) 080406.

[161] Z. Hadzibabic, S. Gupta, C. A. Stan, C. H. Schunck, M. W. Zwierlein, K. Dieckmann, W. Ketterle, Fifty-fold improvement in the number of quantum degenerate fermionic atoms Cond-mat/0306050.

[162] S. R. Granade, M. E. Gehm, K. M. O'Hara, J. E. Thomas, All-optical production of a degenerate Fermi gas, Phys. Rev. Lett. 88 (2002) 120405.

[163] S. Jochim, M. Bartenstein, G. Hendl, J. Hecker-Denschlag, R. Grimm, A. Mosk, M. Weidemüller, Magnetic field control of elastic scattering in a cold gas of fermionic Lithium atoms, Phys. Rev. Lett. 89 (2002) 273202.

[164] K. Dieckmann, C. A. Stan, S. Gupta, Z. Hadzibabic, C. H. Schunck, W. Ketterle, Decay of an ultracold fermionic Lithium gas near a Feshbach resonance, Phys. Rev. Lett. 89 (2002) 203201.

[165] F. Schreck, L. Khaykovich, K. L. Corwin, G. Ferrari, T. Bourdel, J. Cubizolles, C. Salomon, Quasipure Bose-Einstein condensate immersed in a Fermi sea, Phys. Rev. Lett. 87 (2001) 080403.

[166] S. D. Gensemer, D. S. Jin, Transition from collisionless to hydrodynamic behavior in an ultracold Fermi gas, Phys. Rev. Lett. 87 (2001) 173201.

[167] G. Roati, F. Riboli, G. Modugno, M. Inguscio, Fermi-Bose quantum degenerate ^{40}K-^{87}Rb mixture with attractive interaction, Phys. Rev. Lett. 89 (2002) 150403.

[168] J. Goldwin, S. B. Papp, B. DeMarco, D. S. Jin, Two-species magneto-optical trap with ^{40}K and ^{87}Rb, Phys. Rev. A 65 (2002) 021402.

[169] Z. Hadzibabic, C. A. Stan, K. Dieckmann, S. Gupta, M. W. Zwierlein, A. Görlitz, W. Ketterle, Two-species mixture of quantum degenerate Bose and Fermi gases, Phys. Rev. Lett. 88 (2002) 160401.

[170] S. L. Shapiro, S. A. Teukolsky, Black Holes, White Dwarfs and Neutron Stars, the Physics of Compact Objects, Wiley, New York, 1983.

[171] M. Amoruso, I. Meccoli, A. Minguzzi, M. P. Tosi, Density profiles and collective excitations of a trapped two-component Fermi vapor, Eur. Phys. J. D 8 (2000) 361.

[172] L. Vichi, S. Stringari, Collective oscillations of an interacting trapped Fermi gas, Phys. Rev. A 60 (1999) 4734.

[173] B. DeMarco, D. S. Jin, Spin excitations in a Fermi gas of atoms, Phys. Rev. Lett. 88 (2002) 040405.

[174] B. DeMarco, S. B. Papp, D. S. Jin, Pauli blocking of collisions in a quantum degenerate atomic Fermi gas, Phys. Rev. Lett. 86 (2001) 5409.

[175] G. Modugno, G. Roati, F. Riboli, F. Ferlaino, R. J. Brecha, M. Inguscio, Collapse of a degenerate Fermi gas, Science 297 (2002) 2240.

[176] M. Modugno, F. Ferlaino, F. Riboli, G. Roati, G. Modugno, M. Inguscio, Mean-field analysis of the stability of a K-Rb Fermi-Bose mixture, Phys. Rev. A 68 (2003) 043626.

[177] Z. Akdeniz, P. Vignolo, A. Minguzzi, M. P. Tosi, Temperature dependence of density profiles for a cloud of noninteracting fermions moving inside a harmonic trap in one dimension, Phys. Rev. A 66 (2002) 055601.

[178] M. Houbiers, R. Ferwerda, H. T. C. Stoof, W. I. McAlexander, C. A. Sackett, R. G. Hulet, Superfluid state of atomic ^{6}Li in a magnetic trap, Phys. Rev. A 56 (1997) 4864.

[179] M. A. Baranov, Y. Kagan, M. Y. Kagan, On the possibility of a superfluid transition in a Fermi gas of neutral particles at ultralow temperatures, Sov. Phys. JETP Lett. 64 (1996) 301.

[180] D. V. Efremov, L. Viverit, p-wave Cooper pairing of fermions in mixtures of dilute Fermi and Bose gases, Phys. Rev. B 65 (2002) 134519.

[181] H. Heiselberg, C. Pethick, H. Smith, L. Viverit, Influence of induced interactions on the superfluid transition in dilute Fermi gases, Phys. Rev. Lett. 85 (2000) 2418.

[182] M. J. Bijlsma, B. A. Heringa, H. T. C. Stoof, Phonon exchange in dilute Fermi-Bose mixtures: Tailoring the Fermi-Fermi interaction, Phys. Rev. A 61 (2000) 053601.

[183] P. Nozières, S. Schmitt-Rink, Bose-condensation in an attractive fermion gas: from weak to strong coupling superconductivity, J. Low Temp. Phys. 59 (1985) 195.

[184] C. A. R. Sá de Melo, M. Randeria, J. R. Engelbrecht, Crossover from BCS to Bose superconductivity: Transition temperature and time-dependent Ginzburg-Landau theory, Phys. Rev. Lett. 71 (1993) 3202.

[185] Y. Ohashi, A. Griffin, BCS-BEC crossover in a gas of Fermi atoms with a Feshbach resonance, Phys. Rev. Lett. 89 (2002) 130402.

[186] P. Törmä, P. Zoller, Laser probing of atomic Cooper pairs, Phys. Rev. Lett. 85 (2000) 487.

[187] M. A. Baranov, D. S. Petrov, Low-energy collective excitations in a superfluid trapped Fermi gas, Phys. Rev. A 62 (2000) 041601.

[188] G. M. Bruun, B. R. Mottelson, Low energy collective modes of a superfluid trapped atomic Fermi gas, Phys. Rev. Lett. 87 (2001) 270403.

[189] A. Minguzzi, M. P. Tosi, Scissors mode in a superfluid Fermi gas, Phys. Rev. A 63 (2001) 023609.

[190] M. Farine, P. Schuck, X. Vinas, Moment of inertia of a trapped superfluid gas of atomic fermions, Phys. Rev. A 62 (2000) 013608.

[191] A. Minguzzi, G. Ferrari, Y. Castin, Dynamic structure factor of a superfluid Fermi gas, Eur. Phys. J. D 17 (2001) 49.

[192] C. Menotti, P. Pedri, S. Stringari, Expansion of an interacting Fermi gas, Phys. Rev. Lett. 89 (2002) 250402.

[193] A. Perali, P. Pieri, G. C. Strinati, Shrinking of a condensed fermionic cloud in a trap approaching the Bose-Einstein condensation limit, Phys. Rev. A 68 (2003) 031601.

[194] H. Feshbach, A unified theory of nuclear reactions, Ann. Phys. (NY) 19 (1962) 287.

[195] W. C. Stwalley, Stability of spin-aligned Hydrogen at low temperatures and high magnetic fields: New field-dependent scattering resonances and predissociations, Phys. Rev. Lett. 37 (1976) 1628.

[196] E. Tiesinga, B. J. Verhaar, H. T. C. Stoof, Threshold and resonance phenomena in ultracold ground-state collisions, Phys. Rev. A 47 (1993) 4114.

[197] S. Inouye, M. R. Andrews, J. Stenger, H. J. Miesner, D. M. Stamper-Kurn, W. Ketterle, Observation of Feshbach resonances in a Bose-Einstein condensate, Nature 392 (1998) 151.

[198] P. Courteille, R. S. Freeland, D. J. Heinzen, F. A. van Abeelen, B. J. Verhaar, Observation of a Feshbach resonance in cold atom scattering, Phys. Rev. Lett. 81 (1998) 69.

[199] J. L. Roberts, N. R. Claussen, J. P. Burke, Jr., C. H. Greene, E. A. Cornell, C. E. Wieman, Resonant magnetic field control of elastic scattering of cold ^{85}Rb, Phys. Rev. Lett. 81 (1998) 5109.

[200] V. Vuletić, A. J. Kerman, C. Chin, S. Chu, Observation of low-field Feshbach resonances in collisions of Cesium atoms, Phys. Rev. Lett. 82 (1999) 1406.

[201] K. E. Strecker, G. P. Partridge, A. G. Truscott, R. G. Hulet, Formation and propagation of matter-wave soliton trains, Nature 417 (2002) 150.

[202] L. Khaykovich, F. Schreck, G. Ferrari, T. Bourdel, J. Cubizolles, L. D. Carr, Y. Castin, C. Salomon, Formation of a matter-wave bright soliton, Science 296 (2002) 1290.

[203] E. A. Donley, N. R. Claussen, S. T. Thompson, C. E. Wieman, Atom-molecule coherence in a Bose-Einstein condensate, Nature 417 (2002) 529.

[204] P. D. Drummond, K. V. Kheruntsyan, D. J. Heinzen, R. H. Wynar, Stimulated Raman adiabatic passage from an atomic to a molecular Bose-Einstein condensate, Phys. Rev. A 65 (2002) 063619.

[205] T. Loftus, C. A. Regal, C. Ticknor, J. L. Bohn, D. S. Jin, Resonant control of elastic collisions in an optically trapped Fermi gas of atoms, Phys. Rev. Lett. 88 (2002) 173201.

[206] C. A. Regal, C. Ticknor, J. L. Bohn, D. S. Jin, Tuning p-wave interactions in an ultracold Fermi gas of atoms, Phys. Rev. Lett. 90 (2003) 053201.

[207] K. M. O'Hara, S. L. Hemmer, S. R. Granade, M. E. Gehm, J. E. Thomas, V. Venturi, E. Tiesinga, C. J. Williams, Measurement of the zero crossing in a Feshbach resonance of fermionic ^{6}Li, Phys. Rev. A 66 (2002) 041401.

[208] C. A. Regal, D. S. Jin, Measurement of positive and negative scattering lengths in a Fermi gas of atoms, Phys. Rev. Lett. 90 (2003) 230404.

[209] C. A. Regal, C. Ticknor, J. L. Bohn, D. S. Jin, Creation of ultracold molecules from a Fermi gas of atoms, Nature 424 (2003) 47.

[210] J. Cubizolles, T. Bourdel, S. J. J. M. F. Kokkelmans, G. V. Shlyapnikov, C. Salomon, Production of long-lived ultracold Li$_2$ molecules from a Fermi gas Cond-mat/0308018.

[211] L. Carr, G. Shlyapnikov, Y. Castin, Achieving a BCS transition in an atomic Fermi gas Cond-mat/0308036.

[212] M. Greiner, C. A. Regal, D. S. Jin, A molecular Bose-Einstein condensate emerges from a Fermi sea Cond-mat/0311172.

[213] S. Jochim, M. Bartenstein, A. Altmeyer, G. Hendl, S. Riedl, C. Chin, J. H. Denschlag, R. Grimm, Bose-Einstein condensation of molecules, Science Express Science.1093280.

[214] M. W. Zwierlein, C. A. Stan, C. H. Schunck, S. M. F. Raupach, S. Gupta, Z. Hadzibabic, W. Ketterle, Observation of Bose-Einstein condensation of molecules, Phys. Rev. Lett. 91 (2003) 250401.

[215] K. M. O'Hara, S. L. Hemmer, M. E. Gehm, S. R. Granade, J. E. Thomas, Observation of a strongly interacting degenerate Fermi gas of atoms, Science 298 (2002) 2179.

[216] Y. Kagan, E. L. Surkov, G. V. Shlyapnikov, Evolution of a Bose gas in anisotropic time-dependent traps, Phys. Rev. A 55 (1997) R18.

[217] M. E. Gehm, S. L. Hemmer, S. R. Granade, K. M. O'Hara, J. E. Thomas, Mechanical stability of a strongly interacting Fermi gas of atoms, Phys. Rev. A 68 (2003) 011401.

[218] H. Heiselberg, Fermi systems with long scattering lengths, Phys. Rev. A 63 (2001) 043606.

[219] P.-O. Löwdin, Quantum theory of many-particle systems. I. Physical interpretations by means of density matrices, natural spin-orbitals, and convergence problems in the method of configurational interaction, Phys. Rev. 97 (1955) 1474.

[220] E. P. Gross, Structure of a quantized vortex in boson systems, Nuovo Cimento 20 (1961) 451.

[221] E. P. Gross, Hydrodynamics of a superfluid condensate, J. Math. Phys. 4 (1963) 195.

[222] L. P. Pitaevskii, Vortex lines in an imperfect Bose gas, Sov. Phys. JETP 13 (1961) 451.

[223] E. B. Kolomeisky, T. J. Newman, J. P. Straley, X. Qi, Low-dimensional Bose liquids: Beyond the Gross-Pitaevskii approximation, Phys. Rev. Lett. 85 (2000) 1146.

[224] G. Baym, C. J. Pethick, Ground-state properties of magnetically trapped Bose-condensed Rubidium gas, Phys. Rev. Lett. 76 (1996) 6.

[225] L. S. Cederbaum, A. I. Streltsov, Best mean-field for condensates, Phys. Lett. A 318 (2003) 564.

[226] P. Nozières, D. Saint James, Particle vs pair condensation in attractive Bose liquids, J. Phys. 43 (1982) 1133.

[227] G. Baym, Topics in the microscopic theory of Bose-Einstein condensates, J. Phys. B 34 (2001) 4541.

[228] M. D. Girardeau, E. M. Wright, Bose-Fermi variational theory of the Bose-Einstein condensate crossover to the Tonks gas, Phys. Rev. Lett. 87 (2001) 210401.

[229] J. Thijssen, Computational Physics, Cambridge University Press, Cambridge, 1999.

[230] W. Bao, W. Tang, Ground-state solution of Bose-Einstein condensate by directly minimizing the energy functional, J. Comp. Phys. 187 (2003) 230.

[231] B. Dacorogna, Direct Methods in the Calculus of Variations, Vol. 78, Springer, Berlin, 1989, A functional $F(\phi)$ is said to be coercive if it obeys the property $|F(\phi)| \geq a||\phi|| + b$ for each ϕ with $a > 0$ and $b \in R$. Here $||...||$ denotes the norm in the corresponding functional space.

[232] F. Troyon, R. Gruber, H. Saurenmann, S. Semenzato, S. Succi, MHD-limits to plasma confinement, Plasma Phys. Contr. Fusion 26 (1984) 209.

[233] M. Edwards, K. Burnett, Numerical solution of the nonlinear Schrödinger equation for small samples of trapped neutral atoms, Phys. Rev. A 51 (1995) 1382.

[234] S. K. Adhikari, Numerical study of the coupled time-dependent Gross-Pitaevskii equation: Application to Bose-Einstein condensation, Phys. Rev. E 63 (2001) 056704.

[235] W. H. Press, B. P. Flannery, W. T. Vetterling, S. A. Teukolosky, Numerical Recipes: The Art of Scientific Computing, Cambridge University Press, Cambridge, 1992.

[236] M. L. Chiofalo, S. Succi, M. P. Tosi, Ground state of trapped interacting Bose-Einstein condensates by an explicit imaginary-time algorithm, Phys. Rev. E 62 (2000) 7438.

[237] F. Dalfovo, S. Stringari, Bosons in anisotropic traps: Ground state and vortices, Phys. Rev. A 53 (1996) 2477.

[238] E. Lundh, C. J. Pethick, H. Smith, Vortices in Bose-Einstein-condensed atomic clouds, Phys. Rev. A 58 (1998) 4816.

[239] E. L. Bolda, D. F. Walls, Detection of vorticity in Bose-Einstein condensed gases by matter-wave interference, Phys. Rev. Lett. 81 (1998) 5477.

[240] Y. Castin, R. Dum, Bose-Einstein condensates with vortices in rotating traps, Eur. Phys. J. D 7 (1999) 399.

[241] D. S. Rokhsar, Vortex stability and persistent currents in trapped Bose gases, Phys. Rev. Lett. 79 (1997) 2164.

[242] A. A. Svidzinsky, A. L. Fetter, Stability of a vortex in a trapped Bose-Einstein condensate, Phys. Rev. Lett. 84 (2000) 5919.

[243] D. L. Feder, C. W. Clark, B. I. Schneider, Vortex stability of interacting Bose-Einstein condensates confined in anisotropic harmonic traps, Phys. Rev. Lett. 82 (1999) 4956.

[244] D. A. Butts, D. S. Rokhsar, Predicted signatures of rotating Bose-Einstein condensates, Nature 397 (1999) 327.

[245] J. J. García-Ripoll, V. M. Pérez-García, Vortex nucleation and hysteresis phenomena in rotating Bose-Einstein condensates, Phys. Rev. A 63 (2001) 041603(R).

[246] J. J. García-Ripoll, V. M. Pérez-García, Vortex bending and tightly packed vortex lattices in Bose-Einstein condensates, Phys. Rev. A 64 (2001) 053611.

[247] A. Aftalion, T. Riviere, Vortex energy and vortex bending for a rotating Bose-Einstein condensate, Phys. Rev. A 64 (2001) 043611.

[248] M. Modugno, L. Pricoupenko, Y. Castin, Bose-Einstein condensates with a bent vortex in rotating traps, Eur. Phys. J. D 22 (2003) 235.

[249] P. Rosenbusch, V. Bretin, J. Dalibard, Dynamics of a single vortex line in a Bose-Einstein condensate, Phys. Rev. Lett. 89 (2002) 200403.

[250] D. Landau, K. Binder, A Guide to Monte Carlo Simulations in Statistical Physics, Cambridge University Press, Cambridge, 2001.

[251] W. M. C. Foulkes, L. Mitas, R. J. Needs, G. Rajagopal, Quantum Monte Carlo simulations of solids, Rev. Mod. Phys. 73 (2001) 33.

[252] S. Giorgini, J. Boronat, J. Casulleras, A Diffusion Monte Carlo study of a dilute Bose gas, Phys. Rev. A 60 (1999) 5129.

[253] D. Blume, C. H. Greene, Quantum corrections to the ground state of a trapped Bose-Einstein condensate: A Diffusion Monte Carlo calculation, Phys. Rev. A 63 (2001) 063601.

[254] T. D. Lee, K. Huang, C. N. Yang, Eigenvalues and eigenfunctions of a Bose system of hard spheres and its low-temperature properties, Phys. Rev. 106 (1957) 1135.

[255] S. Cowell, H. Heiselberg, I. E. Mazets, J. Morales, V. R. Pandharipande, C. J. Pethick, Cold Bose gases with large scattering lengths, Phys. Rev. Lett. 88 (2002) 210403.

[256] E. Braaten, H.-W. Hammer, M. Kusunoki, Efimov states in a Bose-Einstein condensate near a Feshbach resonance, Phys. Rev. Lett. 90 (2003) 170402.

[257] E. Braaten, H.-W. Hammer, T. Mehen, Dilute Bose-Einstein condensate with large scattering length, Phys. Rev. Lett. 88 (2002) 040401.

[258] G. C. Batrouni, V. Rousseau, R. T. Scalettar, M. Rigol, A. Muramatsu, P. J. H. Denteneer, M. Troyer, Mott domains in confined optical lattices, Phys. Rev. Lett. 89 (2002) 117203.

[259] J. D. Reppy, Superfluid Helium in porous media, J. Low Temp. Phys. 87 (1992) 205.

[260] K. Huang, H. F. Meng, Hard-sphere Bose gas in random external potentials, Phys. Rev. Lett. 69 (1992) 644.

[261] S. Giorgini, L. Pitaevskii, S. Stringari, Effects of disorder in a dilute Bose gas, Phys. Rev. B 49 (1994) 12938.

[262] G. E. Astrakharchik, J. Boronat, J. Casulleras, S. Giorgini, Superfluidity vs. Bose-Einstein condensation in a Bose gas with disorder, Phys. Rev. A 66 (2002) 023603.

[263] J. Boronat, J. Casulleras, J. Navarro, Monte Carlo calculations for liquid ^{4}He at negative pressure, Phys. Rev. B 50 (1994) 3427.

[264] M. C. Gordillo, D. M. Ceperley, Effect of disorder on the critical temperature of a dilute hard-sphere gas, Phys. Rev. Lett. 85 (2000) 4735.

[265] J. O. Andersen, E. Braaten, Semiclassical corrections to a static Bose-Einstein condensate at zero temperature, Phys. Rev. A 60 (1999) 2330.

[266] J. L. DuBois, H. R. Glyde, Bose-Einstein condensation in trapped bosons: A variational Monte Carlo analysis, Phys. Rev. A 63 (2001) 023602.

[267] A. R. Sakhel, J. L. DuBois, H. R. Glyde, Bose-Einstein condensates in ^{85}Rb gases at higher densities, Phys. Rev. A 66 (2002) 063610.

[268] J. L. DuBois, H. R. Glyde, Natural orbitals and Bose-Einstein condensates in traps: A Diffusion Monte Carlo analysis, Phys. Rev. A 68 (2003) 033602.

[269] A. Fabrocini, A. Polls, Bose-Einstein condensates in the large-gas-parameter regime, Phys. Rev. A 64 (2001) 063610.

[270] D. S. Lewart, V. R. Pandharipande, S. C. Pieper, Single-particle orbitals in liquid-Helium drops, Phys. Rev. B 37 (1988) 4950.

[271] A. Fabrocini, A. Polls, Beyond the Gross-Pitaevskii approximation: Local density versus correlated basis approach for trapped bosons, Phys. Rev. A 60 (1999) 2319.

[272] F. Mazzanti, A. Polls, A. Fabrocini, Energy and structure of dilute hard- and soft-sphere gases, Phys. Rev. A 67 (2003) 063615.

[273] L. Tonks, The complete equation of state of one, two and three-dimensional gases of hard elastic spheres, Phys. Rev. 50 (1936) 955.

[274] M. D. Girardeau, Relationship between systems of impenetrable bosons and fermions in one dimension, J. Math. Phys. 1 (1960) 516.

[275] M. D. Girardeau, Permutation symmetry of many-particle wave functions, Phys. Rev. 139 (1965) B500.

[276] A. Lenard, One-dimensional impenetrable bosons in thermal equilibrium, J. Math. Phys. 7 (1966) 1268.

[277] H. Moritz, T. Stöferle, M. Köhl, T. Esslinger, Exciting collective oscillations in a trapped 1D gas Cond-mat/0307607.

[278] E. H. Lieb, W. Liniger, Exact analysis of an interacting Bose gas. I. The general solution and the ground state, Phys. Rev. 130 (1963) 1605.

[279] D. S. Petrov, G. V. Shlyapnikov, J. T. M. Walraven, Regimes of quantum degeneracy in trapped 1D gases, Phys. Rev. Lett. 85 (2000) 3745.

[280] V. Dunjko, V. Lorent, M. Olshanii, Bosons in cigar-shaped traps: Thomas-Fermi regime, Tonks-Girardeau regime, and in between, Phys. Rev. Lett. 86 (2001) 5413.

[281] D. Blume, Fermionization of a bosonic gas under highly elongated confinement: A Diffusion Quantum Monte Carlo study, Phys. Rev. A 66 (2002) 053613.

[282] G. E. Astrakharchik, S. Giorgini, Quantum Monte Carlo study of the three- to one-dimensional crossover for a trapped Bose gas, Phys. Rev. A 66 (2002) 053614.

[283] G. J. Lapeyre Jr., M. D. Girardeau, E. M. Wright, Momentum distribution for a one-dimensional trapped gas of hard-core bosons, Phys. Rev. A 66 (2002) 023606.

[284] G. E. Astrakharchik, S. Giorgini, Correlation functions and momentum distribution of one-dimensional Bose systems, Phys. Rev. A 68 (2003) 031602.

[285] A. Minguzzi, P. Vignolo, M. P. Tosi, High momentum tail in the Tonks gas under harmonic confinement, Phys. Lett. A 294 (2002) 222.

[286] S. Sachdev, Quantum Phase Transitions, Cambridge University Press, Cambridge, 1999.

[287] S. van Oosten, P. van der Straten, H. T. C. Stoof, Quantum phases in an optical lattice, Phys. Rev. A 63 (2001) 053601.

[288] A. M. Rey, K. Burnett, R. Roth, M. Edwards, C. J. Williams, C. W. Clark, Bogoliubov approach to superfluidity of atoms in an optical lattice, J. Phys. B 36 (2003) 825.

[289] H. P. Büchler, G. Blatter, W. Zwerger, Commensurate-incommensurate transition of cold atoms in optical lattices, Phys. Rev. Lett. 90 (2003) 130401.

[290] R. Roth, K. Burnett, Superfluidity and interference pattern of ultracold bosons in optical lattices, Phys. Rev. A 67 (2003) 031602.

[291] V. A. Kashurnikov, N. V. Prokof'ev, B. V. Svistunov, Revealing superfluid-Mott insulator transition in an optical lattice, Phys. Rev. A 66 (2002) 031601.

[292] J. Tobochnik, G. C. Batrouni, H. Gould, Quantum Monte Carlo on a lattice, Comp. Phys. 6 (1992) 673.

[293] M. L. Chiofalo, A. Minguzzi, M. P. Tosi, Time-dependent linear response of an inhomogeneous Bose superfluid: Microscopic theory and connection to current-density functional theory, Physica B 254 (1998) 188.

[294] D. Potter, Computational Physics, Wiley, London, 1973.

[295] K. C. Kulander (Ed.), Time-dependent Methods for Quantum Mechanics, Comput. Phys. Commun. 63 (1991) 1.

159

[296] P. Visscher, A fast explicit algorithm for the time-dependent Schrödinger equation, Comp. Phys. 5 (6) (1991) 596.

[297] D. Frenkel, B. Smit, Understanding Molecular Simulations, Academic, San Diego, 2001.

[298] R. Kosloff, Time-dependent quantum-mechanical methods for molecular-dynamics, J. Phys. Chem. 92 (1988) 2087.

[299] M. M. Cerimele, F. Pistella, M. L. Chiofalo, S. Succi, M. P. Tosi, Numerical solution of the Gross-Pitaevskii equation using an explicit finite-difference scheme: An application to Bose-Einstein condensates, Phys. Rev. E 62 (2000) 1382.

[300] R. Benzi, S. Succi, M. Vergassola, The lattice Boltzmann equation: Theory and applications, Phys. Rep. 222 (1992) 145.

[301] S. Succi, R. Benzi, Lattice Boltzmann equation for quantum mechanics, Physica D 69 (1993) 327.

[302] S. Succi, Numerical solution of the Schrödinger equation using discrete kinetic theory, Phys. Rev. E. 63 (1996) 1969.

[303] S. Succi, Lattice quantum mechanics: An application to Bose-Einstein condensation, Int. J. Mod. Phys. C 9 (1998) 1557.

[304] S. K. Adhikari, Numerical study of the spherically symmetric Gross-Pitaevskii equation in two space dimensions, Phys. Rev. E 62 (2000) 2937.

[305] E. Cerboneschi, R. Mannella, E. Arimondo, L. Salasnich, Oscillation frequencies for a Bose condensate in a triaxial magnetic trap, Phys. Lett. A 249 (1998) 495.

[306] L. Salasnich, A. Parola, L. Reatto, Pulsed macroscopic quantum tunneling of falling Bose-Einstein condensates, Phys. Rev. A 64 (2001) 023601.

[307] P. Muruganandam, S. K. Adhikari, Bose-Einstein condensation dynamics in three dimensions by pseudo-spectral and finite-difference methods, J. Phys. B 36 (2003) 2501.

[308] W. Bao, D. Jaksch, P. Markowich, Numerical solution of the Gross-Pitaevskii equation for Bose-Einstein condensation, J. Comp. Phys. 187 (2003) 318.

[309] M. Edwards, P. A. Ruprecht, K. Burnett, R. J. Dodd, C. W. Clark, Collective excitations of atomic Bose-Einstein condensates, Phys. Rev. Lett. 77 (1996) 1671.

[310] V. M. Pérez-García, H. Michinel, J. I. Cirac, M. Lewenstein, P. Zoller, Low energy excitations of a Bose-Einstein condensate: A time-dependent variational analysis, Phys. Rev. Lett. 77 (1996) 5320.

[311] S. Stringari, Collective excitations of a trapped Bose-condensed gas, Phys. Rev. Lett. 77 (1996) 2360.

160

[312] H. Wallis, A. Röhrl, M. Naraschewski, A. Schenzle, Phase-space dynamics of Bose condensates: Interference *versus* interaction, Phys. Rev. A 55 (1997) 2109.

[313] M. L. Chiofalo, S. Succi, M. P. Tosi, Output coupling of Bose condensates from atomic tunnel arrays: A numerical study, Phys. Lett. A 260 (1999) 86.

[314] M. L. Chiofalo, M. Polini, M. P. Tosi, Coherent transport in a Bose-Einstein condensate inside an optical lattice, Laser Phys. 12 (2002) 50.

[315] M. Cardenas, M. L. Chiofalo, M. P. Tosi, Matter wave dynamics in an optical lattice: Decoherence of Josephson-type oscillations from the Gross-Pitaevskii equation, Physica B 322 (2002) 116.

[316] S. A. Gardiner, D. Jaksch, R. Dum, J. I. Cirac, P. Zoller, Nonlinear matter wave dynamics with a chaotic potential, Phys. Rev. A 62 (2000) 023612.

[317] M. Polini, R. Fazio, M. P. Tosi, J. Sinova, A. H. MacDonald, Frustration of a Bose gas inside an optical lattice, Laser Phys. (in press) .

[318] A. Recati, F. Zambelli, S. Stringari, Overcritical rotation of a trapped Bose-Einstein condensate, Phys. Rev. Lett. 86 (2001) 377.

[319] S. Sinha, Y. Castin, Dynamic instability of a rotating Bose-Einstein condensate, Phys. Rev. Lett. 87 (2001) 190402.

[320] F. Dalfovo, S. Stringari, Shape deformations and angular-momentum transfer in trapped Bose-Einstein condensates, Phys. Rev. A 63 (2001) 011601.

[321] M. Tsubota, K. Kasamatsu, M. Ueda, Vortex lattice formation in a rotating Bose-Einstein condensate, Phys. Rev. A 65 (2002) 023603.

[322] A. A. Penckwitt, R. J. Ballagh, C. W. Gardiner, Nucleation, growth, and stabilization of Bose-Einstein condensate vortex lattices, Phys. Rev. Lett. 89 (2002) 260402.

[323] C. Lobo, A. Sinatra, Y. Castin, Vortex crystallization in classical field theory, Phys. Rev. Lett. 92 (2004) 020403.

[324] B. Jackson, J. F. McCann, C. S. Adams, Vortex formation in dilute inhomogeneous Bose-Einstein condensates, Phys. Rev. Lett. 80 (1998) 3903.

[325] V. Bretin, P. Rosenbusch, F. Chevy, G. V. Shlyapnikov, J. Dalibard, Quadrupole oscillation of a single-vortex Bose-Einstein condensate: Evidence for Kelvin modes, Phys. Rev. Lett. 90 (2003) 100403.

[326] T. Mizushima, M. Ichioka, K. Machida, Beliaev damping and Kelvin mode spectroscopy of a Bose-Einstein condensate in the presence of a vortex line, Phys. Rev. Lett. 90 (2003) 180401.

[327] L. O. Baksmaty, S. J. Woo, S. Choi, N. P. Bigelow, Tkachenko waves in rapidly rotating Bose-Einstein condensates Cond-mat/0307368.

[328] I. Coddington, P. Engels, V. Schweikhard, E. A. Cornell, Observation of Tkachenko oscillations in rapidly rotating Bose-Einstein condensates, Phys. Rev. Lett. 91 (2003) 100402.

[329] G. Baym, Tkachenko modes of vortex lattices in rapidly rotating Bose-Einstein condensates, Phys. Rev. Lett. 91 (2003) 110402.

[330] Y. Choi, J. Javainen, I. Koltracht, M. Kostrun, M. Kenna, N. Savystka, A fast algorithm for the solution of the time-independent Gross-Pitaevskii equation, J. Comp. Phys. 190 (2003) 1.

[331] P. Engels, I. Coddington, P. C. Haljan, V. Schweikhard, E. A. Cornell, Observation of long-lived vortex aggregates in rapidly rotating Bose-Einstein condensates, Phys. Rev. Lett. 90 (2003) 170405.

[332] T. P. Simula, A. A. Penckwitt, R. J. Ballagh, Giant vortex lattice deformations in rapidly rotating Bose-Einstein condensates Cond-mat/0307130.

[333] P. O. Fedichev, A. E. Muryshev, G. V. Shlyapnikov, Dissipative dynamics of a kink state in a Bose-condensed gas, Phys. Rev. A 60 (1999) 3220.

[334] H. Pu, C. K. Law, J. H. Eberly, N. P. Bigelow, Coherent disintegration and stability of vortices in trapped Bose condensates, Phys. Rev. A 59 (1999) 1533.

[335] J. J. García-Ripoll, G. Molina-Terriza, V. M. Pérez-García, L. Torner, Structural instability of vortices in Bose-Einstein condensates, Phys. Rev. Lett. 87 (2001) 140403.

[336] P. O. Fedichev, G. V. Shlyapnikov, Dissipative dynamics of a vortex state in a trapped Bose-condensed gas, Phys. Rev. A 60 (1999) R1779.

[337] J. E. Williams, M. J. Holland, Preparing topological states of a Bose-Einstein condensate, Nature 401 (1999) 568.

[338] H. Pu, N. P. Bigelow, Properties of two-species Bose condensates, Phys. Rev. Lett. 80 (1998) 1130.

[339] H. Pu, N. P. Bigelow, Collective excitations, metastability, and nonlinear response of a trapped two-species Bose-Einstein condensate, Phys. Rev. Lett. 80 (1998) 1134.

[340] C. Savage, J. Ruostekoski, Energetically stable particle-like skyrmions in a trapped Bose-Einstein condensate, Phys. Rev. Lett. 91 (2003) 010403.

[341] S. Kohler, F. Sols, Oscillatory decay of a two-component Bose-Einstein condensate, Phys. Rev. Lett. 89 (2002) 060403.

[342] J. J. García-Ripoll, V. M. Pérez-García, F. Sols, Split vortices in optically coupled Bose-Einstein condensates, Phys. Rev. A 66 (2002) 021602.

[343] R. A. Battye, N. Cooper, P. Sutcliffe, Stable skyrmions in two-component Bose-Einstein condensates, Phys. Rev. Lett. 88 (2002) 080401.

[344] P. Öheberg, L. Santos, Dark solitons in a two-component Bose-Einstein condensate, Phys. Rev. Lett. 86 (2001) 2918.

[345] M. R. Matthews, B. P. Anderson, P. C. Haljan, D. S. Hall, C. E. Wieman, E. A. Cornell, Watching a superfluid untwist itself: Recurrence of Rabi oscillations in a Bose-Einstein condensate, Phys. Rev. Lett. 83 (1999) 3358.

[346] F. Sols, Josephson effect between Bose condensates, in: M. Inguscio, S. Stringari, C. E. Wieman (Eds.), Bose Einstein condensation in atomic gases, Proc. Int. School "Enrico Fermi", IOP Press, Amsterdam, 1999.

[347] E. M. Lifshitz, L. P. Pitaevskii, Statistical Physics, Part 2, Pergamon, Oxford, 1980.

[348] Y. Kagan, G. V. Shlyapnikov, J. T. M. Walraven, Bose-Einstein condensation in trapped atomic gases, Phys. Rev. Lett. 76 (1996) 2670.

[349] C. C. Bradley, C. A. Sackett, R. G. Hulet, Bose-Einstein condensation of Lithium: Observation of limited condensate number, Phys. Rev. Lett. 78 (1997) 985.

[350] J. L. Roberts, N. R. Claussen, S. L. Cornish, E. A. Donley, E. A. Cornell, C. E. Wieman, Controlled collapse of a Bose-Einstein condensate, Phys. Rev. Lett. 86 (2001) 4211.

[351] P. A. Ruprecht, M. J. Holland, K. Burnett, Time-dependent solution of the nonlinear Schrödinger equation for Bose-condensed trapped neutral atoms, Phys. Rev. A 51 (1995) 4704.

[352] A. Eleftheriou, K. Huang, Instability of a Bose-Einstein condensate with an attractive interaction, Phys. Rev. A 61 (2000) 043601.

[353] C. M. Savage, N. P. Robins, J. J. Hope, Bose-Einstein condensate collapse: A comparison between theory and experiment, Phys. Rev. A 67 (2003) 014304.

[354] L. P. Pitaevskii, Dynamics of collapse of a confined Bose gas, Phys. Lett. A 221 (1996) 14.

[355] E. V. Shuryak, Metastable Bose condensate made of atoms with attractive interaction, Phys. Rev. A 54 (1996) 3151.

[356] H. T. C. Stoof, Macroscopic quantum tunneling of a Bose condensate, J. Stat. Phys. 87 (1997) 1353.

[357] M. Ueda, A. J. Leggett, Macroscopic quantum tunneling of a Bose-Einstein condensate with attractive interaction, Phys. Rev. Lett. 80 (1998) 1576.

[358] Y. Kagan, A. E. Muryshev, G. V. Shlyapnikov, Collapse and Bose-Einstein condensation in a trapped Bose gas with negative scattering length, Phys. Rev. Lett. 81 (1998) 933.

[359] J. M. Gerton, D. Strekalov, I. Prodan, R. G. Hulet, Direct observation of growth and collapse of a Bose-Einstein condensate with attractive interactions, Nature 408 (2000) 692.

[360] E. A. Donley, N. R. Claussen, S. L. Cornish, J. L. Roberts, E. A. Cornell, C. E. Wieman, Dynamics of collapsing and exploding Bose-Einstein condensates, Nature 412 (2001) 295.

[361] B. D. Esry, C. H. Greene, J. P. Burke, Recombination of three atoms in the ultracold limit, Phys. Rev. Lett. 83 (1999) 1751.

[362] H. Saito, M. Ueda, Intermittent implosion and pattern formation of trapped Bose-Einstein condensates with an attractive interaction, Phys. Rev. Lett. 86 (2001) 1406.

[363] U. Al Khawaja, H. T. C. Stoof, R. G. Hulet, K. E. Strecker, G. B. Partridge, Bright soliton trains of trapped Bose-Einstein condensates, Phys. Rev. Lett. 89 (2002) 200404.

[364] L. Salasnich, A. Parola, L. Reatto, Modulational instability and complex dynamics of confined matter-wave solitons, Phys. Rev. Lett. 91 (2003) 080405.

[365] K. Tai, A. Hasegawa, A. Tomita, Observation of modulational instability in optical fibers, Phys. Rev. Lett. 56 (1986) 135.

[366] M. D. Girardeau, E. M. Wright, Breakdown of time-dependent mean-field theory for a one-dimensional condensate of impenetrable bosons, Phys. Rev. Lett. 84 (2000) 5239.

[367] A. Minguzzi, P. Vignolo, M. L. Chiofalo, M. P. Tosi, Hydrodynamic excitations in a spin-polarized Fermi gas under harmonic confinement in one dimension, Phys. Rev. A 64 (2001) 033605.

[368] J. N. Fuchs, X. Leyronas, R. Combescot, Collective modes of a trapped Lieb-Liniger gas: A hydrodynamic approach Cond-mat/0309276.

[369] P. Vignolo, A. Minguzzi, M. P. Tosi, Light scattering from a degenerate quasi-one-dimensional confined gas of non-interacting fermions, Phys. Rev. A 64 (2001) 023421.

[370] A. Csordás, R. Graham, Collective excitations in Bose-Einstein condensates in triaxially anisotropic parabolic traps, Phys. Rev. A 59 (1999) 1477.

[371] S. Stringari, Dynamics of Bose-Einstein condensed gases in highly deformed traps, Phys. Rev. A 58 (1998) 2385.

[372] T.-L. Ho, M. Ma, Quasi 1D and 2D dilute Bose gas in magnetic traps: Existence of off-diagonal order and anomalous quantum fluctuations, J. Low Temp. Phys. 115 (1999) 61.

[373] C. Menotti, S. Stringari, Collective oscillations of a one-dimensional trapped Bose-Einstein gas, Phys. Rev. A 66 (2002) 043610.

[374] A. Minguzzi, S. Conti, M. P. Tosi, The internal energy and condensate fraction of a trapped interacting Bose gas, J. Phys.: Condens. Matter 9 (1997) L33.

[375] F. Dalfovo, S. Giorgini, M. Guilleumas, L. Pitaevskii, S. Stringari, Collective and single-particle excitations of a trapped Bose gas, Phys. Rev. A 56 (5) (1997) 3840.

[376] M. Holzmann, W. Krauth, M. Naraschewski, Precision Monte Carlo test of the Hartree-Fock approximation for a trapped Bose gas, Phys. Rev. A 59 (1999) 2956.

[377] A. Minguzzi, P. Vignolo, M. P. Tosi, Momentum distribution of an interacting Bose-Einstein condensed gas at finite temperature, Phys. Rev. A 62 (2000) 023604.

[378] M. Bayindir, B. Tanatar, Bose-Einstein condensation in a two-dimensional, trapped interacting gas, Phys. Rev. A 58 (1998) 3134.

[379] M. Amoruso, A. Minguzzi, S. Stringari, M. P. Tosi, L.Vichi, Temperature-dependent density profiles of trapped boson-fermion mixtures, Eur. Phys. J. D 4 (1998) 261.

[380] D. M. Ceperley, Path integrals in the theory of condensed Helium, Rev. Mod. Phys. 67 (1995) 279.

[381] W. Krauth, Quantum Monte Carlo calculations for a large number of bosons in a harmonic trap, Phys. Rev. Lett. 77 (1996) 3695.

[382] M. Holzmann, Y. Castin, Pair correlation function of an inhomogeneous interacting Bose-Einstein condensate, Eur. Phys. J. D 7 (2000) 425.

[383] M. Suzuki, Relationship between d-dimensional quantal spin systems and (d+1)-dimensional Ising systems. Equivalence, critical exponents and systematic approximants of the partition function and spin correlations, Prog. Theor. Phys. 56 (1976) 1454.

[384] G. G. Batrouni, R. T. Scalettar, World-line Quantum Monte Carlo algorithm for a one-dimensional Bose model, Phys. Rev. B 46 (1992) 9051.

[385] N. Prokof'ev, B. Svistunov, I. S. Tupitsyn, Exact Quantum Monte Carlo process for the statistics of discrete systems, Pis'ma v Zh.Eks. Teor. Fiz. 64 (1996) 853.

[386] M. Holzmann, W. Krauth, Transition temperature of the homogeneus weakly interacting Bose gas, Phys. Rev. Lett. 83 (1999) 2687.

[387] V. A. Kashurnikov, N. V. Prokof'ev, B. V. Svistunov, Critical temperature shift in weakly interacting Bose gas, Phys. Rev. Lett. 87 (2001) 120402.

[388] J. M. Kosterlitz, D. J. Thouless, Ordering, metastability and phase transition in two-dimensional models, J. Phys. C 6 (1973) 1181.

[389] J. M. Kosterlitz, The critical properties of the two-dimensional XY model, J. Phys. C 7 (1974) 1046.

[390] R. Swendsen, J.-S. Wang, Nonuniversal critical dynamics in Monte Carlo simulations, Phys. Rev. Lett. 58 (1987) 86.

[391] N. V. Prokof'ev, B. V. Svistunov, Worm algorithms for classical statistical mechanics, Phys. Rev. Lett. 87 (2001) 160601.

[392] N. V. Prokof'ev, O. Ruebenacker, B. V. Svistunov, Critical point of a weakly interacting two-dimensional Bose gas, Phys. Rev. Lett. 87 (2001) 270402.

[393] N. V. Prokof'ev, B. V. Svistunov, Two-dimensional weakly interacting Bose gas in the fluctuation region, Phys. Rev. A 66 (2002) 043608.

[394] D. Nelson, J. M. Kosterlitz, Universal jump in the superfluid density of two-dimensional superfluids, Phys. Rev. Lett. 39 (1977) 1201.

[395] R. J. Dodd, M. Edwards, C. W. Clark, K. Burnett, Collective excitations of Bose-Einstein-condensed gas at finite temperature, Phys. Rev. A 57 (1998) R32.

[396] P. O. Fedichev, G. V. Shlyapnikov, Finite-temperature perturbation theory for a spatially inhomogeneous Bose-condensed gas, Phys. Rev. A 58 (1998) 3146.

[397] S. Giorgini, Collisionless dynamics of dilute Bose gases: Role of quantum and thermal fluctuations, Phys. Rev. A 61 (2000) 063615.

[398] J. Reidl, G. Bene, R. Graham, P. Szépfalusy, Kohn mode for trapped Bose gases whithin the dielectric formalism, Phys. Rev. A 63 (2001) 043605.

[399] A. Minguzzi, M. P. Tosi, Linear density response in the random phase approximation for confined Bose vapours at finite temperature, J. Phys: Condens. Matter 9 (1997) 10211.

[400] X.-J. Liu, H. Hu, A. Minguzzi, M. P. Tosi, Collective oscillations of a confined Bose gas at finite temperature in the random phase approximation Cond-mat/0311411.

[401] B. Jackson, E. Zaremba, Finite-temperature simulations of the scissors mode in Bose-Einstein condensed gases, Phys. Rev. Lett. 87 (2001) 100404.

[402] B. Jackson, E. Zaremba, Modeling Bose-Einstein condensed gases at finite temperatures with N-body simulations, Phys. Rev. A 66 (2002) 033606.

[403] P. Vignolo, M. L. Chiofalo, S. Succi, M. P. Tosi, Explicit finite-difference and particles method for the dynamics of mixed Bose-condensate and cold-atom clouds, J. Comp. Phys. 182 (2002) 368.

[404] K. M. Keown, Stochastic Simulation in Physics, Springer, Berlin, 1997.

[405] O. Maragó, G. Hechenblaikner, E. Hodby, C. J. Foot, Temperature dependence on damping and frequency shifts of the scissors mode of a trapped Bose-Einstein condensate, Phys. Rev. Lett. 86 (2001) 3938.

[406] T. Nikuni, Finite-temperature theory of the scissors mode in a Bose gas using the moment method, Phys. Rev. A 65 (2002) 033611.

[407] O. J. Luiten, M. W. Reynolds, J. T. M. Walraven, Kinetic theory of the evaporative cooling of trapped gases, Phys. Rev. A 53 (1996) 381.

[408] E. Mandonnet, A. Minguzzi, R. Dum, I. Carusotto, Y. Castin, J. Dalibard, Evaporative cooling of an atomic beam, Eur. Phys. J. D 10 (2000) 9.

[409] M. J. Davis, S. A. Morgan, K. Burnett, Simulations of Bose fields at finite temperature, Phys. Rev. Lett. 87 (2001) 160402.

[410] H. Umezawa, Advanced Field Theory: Micro, Macro, and Thermal Physics, AIP Press, New York, 1993.

[411] J. R. Dorfman, An Introduction to Chaos in Nonequilibrium Statistical Mechanics, Cambridge University Press, Cambridge, 1999.

[412] A. Sinatra, C. Lobo, Y. Castin, Classical-field method for time-dependent Bose-Einstein condensation, Phys. Rev. Lett. 87 (2001) 210404.

[413] A. Sinatra, C. Lobo, Y. Castin, The truncated Wigner approach or classical field method for Bose-condensed gases: Limits of validity and applications, J. Phys. B 35 (2002) 3599.

[414] I. Carusotto, Y. Castin, J. Dalibard, N-boson time-dependent problem: A reformulation with stochastic wavefunctions, Phys. Rev. A 63 (2001) 023606.

[415] I. Carusotto, Y. Castin, Condensate statistics in one-dimensional interacting Bose gases: Exact results, Phys. Rev. Lett. 90 (2003) 030401.

[416] N. G. Berloff, B. V. Svistunov, Scenario of strongly nonequilibrated Bose-Einstein condensation, Phys. Rev. A 66 (2002) 013603.

[417] R. Jancel, L. de Broglie, Foundations of Classical and Quantum Statistical Mechanics, Pergamon, Oxford, 1969.

[418] D. Jaksch, C. W. Gardiner, P. Zoller, Quantum kinetic theory. II. Simulation of the quantum Boltzmann master equation, Phys. Rev. A 56 (1997) 575.

[419] C. W. Gardiner, P. Zoller, Quantum kinetic theory. III. Quantum kinetic master equation for strongly condensed trapped systems, Phys. Rev. A 58 (1) (1998) 536.

[420] D. Jaksch, C. W. Gardiner, K. M. Gheri, P. Zoller, Quantum kinetic theory. IV. Intensity and amplitude fluctuations of a Bose-Einstein condensate at finite temperature including trap loss, Phys. Rev. A 58 (2) (1998) 1450.

[421] C. W. Gardiner, P. Zoller, Quantum kinetic theory. V. Quantum kinetic master equation for mutual interaction of condensate and noncondensate, Phys. Rev. A 61 (2000) 033601.

[422] M. D. Lee, C. W. Gardiner, Quantum kinetic theory. VI. The growth of a Bose-Einstein condensate, Phys. Rev. A 62 (2000) 033606.

[423] M. J. Davis, C. W. Gardiner, R. J. Ballagh, Quantun kinetic theory. VII. The influence of vapor dynamics on condensate growth, Phys. Rev. A 62 (2000) 063608.

[424] C. W. Gardiner, M. D. Lee, R. J. Ballagh, M. J. Davis, P. Zoller, Quantum kinetic theory of condensate growth: Comparison of experiment and theory, Phys. Rev. Lett. 81 (1998) 5266.

[425] M. Yamashita, M. Koashi, N. Imoto, Quantum kinetic theory for evaporative cooling of trapped atoms: Growth of Bose-Einstein condensate, Phys. Rev. A 59 (1999) 2243.

167

[426] M. J. Bijlsma, E. Zaremba, H. T. C. Stoof, Condensate growth in trapped Bose gases, Phys. Rev. A 62 (2000) 063609.

[427] M. Köhl, M. J. Davis, C. W. Gardiner, T. W. Hänsch, T. Esslinger, Growth of Bose-Einstein condensates from thermal vapor, Phys. Rev. Lett. 88 (2002) 080402.

[428] M. J. Holland, J. Williams, K. Coakley, J. Cooper, Trajectory simulation of kinetic equations for classical systems, Quantum Semiclass. Opt. 8 (1996) 571.

[429] J. Hofbauer, K. Sigmund, Evolutionary Games and Population Dynamics, Cambridge University Press, Cambridge, 1998.

[430] S. Succi, F. Castiglione, M. Bernaschi, Collective dynamics in the immune system response, Phys. Rev. Lett. 79 (1997) 4493.

[431] M. J. Holland, J. Williams, J. Cooper, Bose-Einstein condensation: Kinetic evolution obtained from simulated trajectories, Phys. Rev. A 55 (1997) 3670.

[432] W. Paul, J. Baschnagel, Stochastic Processes from Physics to Finance, Springer, Berlin, 1999.

[433] M. Yamashita, M. Koashi, M. Mitsunaga, N. Imoto, T. Mukai, Optimization of evaporative cooling towards a large number of Bose-Einstein-condensed atoms, Phys. Rev. A 67 (2003) 023601.

[434] W. Geist, L. You, T. A. B. Kennedy, Sympathetic cooling of an atomic Bose-Fermi gas mixture, Phys. Rev. A 59 (1999) 1500.

[435] W. Geist, A. Idrizbegovic, M. Marinescu, T. A. B. Kennedy, L. You, Evaporative cooling of trapped fermionic atoms, Phys. Rev. A 61 (2000) 013406.

[436] M. J. Holland, B. DeMarco, D. S. Jin, Evaporative cooling of a two-component degenerate Fermi gas, Phys. Rev. A 61 (2000) 053610.

[437] W. Geist, T. A. B. Kennedy, Evaporative cooling of mixed atomic fermions, Phys. Rev. A 65 (2002) 063617.

[438] P. Michler, Single Quantum Dots: Fundamentals, Applications, and New Concepts, Springer, Berlin, 2003.

[439] P. B. van Zyl, D. A. W. Hutchinson, Charged two-dimensional quantum gas in a uniform magnetic field at finite temperature, Phys. Rev. B 69.

[440] N. H. March, L. M. Nieto, Analytical relations between kinetic-energy and particle densities for one-dimensional harmonically confined Fermi vapors, Phys. Rev. A 63 (2001) 044502.

[441] M. Brack, B. P. van Zyl, Simple analytical particle and kinetic energy densities for a dilute fermionic gas in a D-dimensional harmonic trap, Phys. Rev. Lett. 86 (2001) 1574.

[442] M. Brack, M. V. N. Murthy, Harmonically trapped fermion gases: Exact and asymptotic results in arbitrary dimensions, J. Phys. A 36 (2003) 1111.

[443] B. P. van Zyl, R. K. Bhaduri, A. Suzuki, M. Brack, Some exact results for a trapped quantum gas at finite temperature, Phys. Rev. A 67 (2003) 023609.

[444] B. P. van Zyl, Exact first-order density matrix for a D-dimensional harmonically confined Fermi gas at finite temperature Cond-mat/0306320.

[445] P. Vignolo, A. Minguzzi, M. P. Tosi, Exact particle and kinetic-energy densities for one-dimensional confined gases of noninteracting fermions, Phys. Rev. Lett. 85 (2000) 2850.

[446] P. Vignolo, A. Minguzzi, M. P. Tosi, Degenerate gases under harmonic confinement in one dimension: Rigorous results in the impenetrable-bosons/spin-polarized fermions limit, J. Mod. Phys. B 16 (2002) 2161.

[447] R. Farchioni, G. Grosso, P. Vignolo, Density of states for energy-dependent effective Hamiltonians, Phys. Rev. B 62 (2000) 12565.

[448] A. Minguzzi, P. Vignolo, M. P. Tosi, Momentum flux density, kinetic energy density, and their fluctuations for one-dimensional confined gases of noninteracting fermions, Phys. Rev. A 63 (2001) 063604.

[449] R. Farchioni, G. Grosso, G. Pastori Parravicini, Renormalization approach for transport and electronic properties of conducting polymers, Phys. Rev. B 53 (1996) 4294.

[450] P. Vignolo, A. Minguzzi, Shell structure in the density profiles for non-interacting fermions in anisotropic harmonic confinement, Phys. Rev. A 67 (2003) 053601.

[451] Z. Akdeniz, P. Vignolo, A. Minguzzi, M. P. Tosi, Temperature dependence of density profiles for a cloud of noninteracting fermions moving inside a harmonic trap in one dimension, Phys. Rev. A 66 (2002) 055601.

[452] L. Wilets, J. Cohen, Fermion molecular dynamics in atomic, molecular, and optical physics, Contemp. Phys. 39 (1998) 163.

[453] H. Feldmeier, J. Schnack, Molecular dynamics for fermions, Rev. Mod. Phys. 72 (2000) 655.

[454] F. Toschi, P. Vignolo, S. Succi, M. P. Tosi, Dynamics of trapped two-component Fermi gas: Temperature dependence of the transition from collisionless to collisional regime, Phys. Rev. A 67 (2003) 041605.

[455] H. Wu, C. J. Foot, Direct simulation of evaporative cooling, J. Phys. B 29 (1996) L321.

[456] T. Nikuni, E. Zaremba, A. Griffin, Two-fluid dynamics for a Bose-Einstein condensate out of local equilibrium with the noncondensate, Phys. Rev. Lett. 83 (1999) 10.

[457] M. Amoruso, I. Meccoli, A. Minguzzi, M. P. Tosi, Collective excitations of a degenerate Fermi vapour in a magnetic trap, Eur. Phys. J. D 7 (1999) 441.

[458] F. Toschi, P. Capuzzi, S. Succi, P. Vignolo, M. P. Tosi, Transition to hydrodynamics in colliding fermion clouds 37 (2004) S91.

[459] P. Nettesheim, F. A. Bornemann, B. Schmidt, C. Schütte, An explicit and symplectic integrator for quantum-classical molecular dynamics, Chem. Phys. Lett. 256 (1996) 581.

[460] S. Succi, F. Toschi, P. Capuzzi, P. Vignolo, M. P. Tosi, A particle-dynamics study of dissipation in colliding clouds of ultracold fermions, Phil. Trans. R. Soc. In press.

[461] Z. Akdeniz, P. Vignolo, M. P. Tosi, Collective dynamics of fermion clouds in cigar-shaped traps, Phys. Lett. A 311 (2003) 246.

[462] K. Mølmer, Bose condensates and Fermi gases at zero temperature, Phys. Rev. Lett. 80 (1998) 1804.

[463] N. Nygaard, K. Mølmer, Component separation in harmonically trapped boson-fermion mixtures, Phys. Rev. A 59 (1999) 2974.

[464] Z. Akdeniz, A. Minguzzi, P. Vignolo, M. P. Tosi, Demixing in mesoscopic boson-fermion clouds inside cylindrical harmonic traps: Quantum phase diagram and role of temperature, Phys. Rev. A 66 (2002) 013620.

[465] Z. Akdeniz, P. Vignolo, A. Minguzzi, M. P. Tosi, Phase separation in a boson-fermion mixture of Lithium atoms, J. Phys. B 35 (2002) L105.

[466] P. Capuzzi, A. Minguzzi, M. P. Tosi, Collective excitations of a trapped boson-fermion mixture across demixing, Phys. Rev. A 67 (2003) 053605.

[467] P. Capuzzi, A. Minguzzi, M. P. Tosi, Collective excitations in trapped boson-fermions mixtures: From demixing to collapse, Phys. Rev. A 68 (2003) 033605.

[468] T. Miyakawa, T. Suzuki, H. Yabu, Sum-rule approach to collective oscillations of a boson-fermion mixed condensate of alkali atoms, Phys. Rev. A 62 (2000) 063613.

[469] X.-J. Liu, H. Hu, Collisionless and hydrodynamic excitations of trapped boson-fermion mixtures, Phys. Rev. A 67 (2003) 023613.

[470] P. Capuzzi, E. S. Hernández, Zero-sound density oscillations in Fermi-Bose mixtures, Phys. Rev. A 64 (2001) 043607.

[471] T. Sogo, T. Miyakawa, T. Suzuki, H. Yabu, Random-phase approximation study of collective excitations in the Bose-Fermi mixed condensate of alkali-metal gases, Phys. Rev. A 66 (2002) 013618.

[472] C. F. von Weizsäcker, Zur theorie der kernmassen, Z. Phys. 96 (1935) 431.

[473] P. Capuzzi, A. Minguzzi, M. P. Tosi, Dynamic *versus* thermodynamic signatures of quantum demixing under confinement, Laser Phys. In press.

[474] T. Miyakawa, T. Suzuki, H. Yabu, Induced instability for boson-fermion mixed condensate of alkali atoms due to attractive boson-fermion interaction, Phys. Rev. A 64 (2001) 033611.

[475] R. Roth, H. Feldmeier, Mean-field instability of trapped dilute boson-fermion mixtures, Phys. Rev. A 65 (2002) 021603.

[476] R. Roth, Structure and stability of trapped atomic boson-fermion mixtures, Phys. Rev. A 66 (2002) 013614.

[477] L. Viverit, C. J. Pethick, H. Smith, Zero-temperature phase diagram of binary boson-fermion mixtures, Phys. Rev. A 61 (2000) 053605.

[478] A. Minguzzi, M. P. Tosi, Schematic phase diagram and collective excitations in the collisional regime for trapped boson-fermion mixtures at zero temperature, Phys. Lett. A 268 (2000) 142.

[479] Z. Akdeniz, A. Minguzzi, P. Vignolo, Demixing of boson-fermion clouds under harmonic confinement, Laser Phys. 13 (2003) 577.

[480] F. Ferlaino, R. J. Brecha, P. Hannaford, F. Riboli, G. Roati, G. Modugno, M. Inguscio, Dipolar oscillations in a quantum degenerate Fermi-Bose atomic mixture, J. Opt. B 5 (2003) S3.

[481] N. H. March, M. P. Tosi, Quantum theory of pure liquid metals as two-component systems, Ann. Phys. (NY) 81 (1973) 414.

[482] R. J. Dodd, M. Edwards, C. J. Williams, C. W. Clark, M. J. Holland, P. A. Ruprecht, K. Burnett, Role of attractive interactions on Bose-Einstein condensation, Phys. Rev. A 54 (1996) 661.

[483] P. Capuzzi, A. Minguzzi, M. P. Tosi, Excitation spectra of a confined ^{87}Rb-^{40}K mixture approaching collapse, J. Phys. B 37 (2004) S73.

[484] H. Hu, X.-J. Liu, Thermodynamics of a trapped Bose-Fermi mixture, Phys. Rev. A 68 (2003) 023608.

[485] P. Capuzzi, A. Minguzzi, M. P. Tosi, Collisional oscillations of trapped boson-fermion mixtures approaching collapse Cond-mat/0401331.

[486] A. Perali, P. Pieri, G. C. Strinati, C. Castellani, Pseudogap and spectral function from superconducting fluctuations to the bosonoc limit, Phys. Rev. B 66 (2002) 024510.

[487] R. Micnas, J. Ranninger, S. Robaszkiewicz, Superconductivity in narrow-band systems with local nonretarded attractive interactions, Rev. Mod. Phys. 62 (1990) 113.

[488] W. Hofstetter, J. I. Cirac, P. Zoller, E. Demler, M. D. Lukin, High-temperature superfluidity of fermionic atoms in optical lattices, Phys. Rev. Lett. 89 (2002) 220407.

[489] M. Rodríguez, P. Törmä, Bloch oscillations in Fermi gases Cond-mat/0303634.

[490] M. Holland, S. J. J. M. F. Kokkelmans, M. L. Chiofalo, R. Walser, Resonance superfluidity in a quantum degenerate Fermi gas, Phys. Rev. Lett. 87 (2001) 120406.

[491] E. Timmermans, P. Tommasini, M. Hussein, A. K. Kerman, Feshbach resonances in atomic Bose-Einstein condensates, Phys. Rep. 315 (1999) 199.

[492] M. L. Chiofalo, S. J. J. M. F. Kokkelmans, J. N. Milstein, M. J. Holland, Signatures of resonance superfluidity in a quantum Fermi gas, Phys. Rev. Lett. 88 (2002) 090402.

[493] Y. Ohashi, A. Griffin, Superfluid transition temperature in a trapped gas of Fermi atoms with a Feshbach resonance, Phys. Rev. A 67 (2003) 033603.

[494] J. Ranninger, S. Robaszkiewicz, Superconductivity of locally paired electrons, Physica B 135 (1985) 468.

[495] M. Marini, F. Pistolesi, G. C. Strinati, Evolution from BCS superconductivity to Bose condensation: Analytic results for the crossover in three dimensions, Eur. Phys. J. B 1 (1998) 151.

[496] T. Papenbrock, G. F. Bertsch, Pairing in low-density Fermi gases, Phys. Rev. C 59 (1999) 2052.

[497] J. Carlson, S.-Y. Chang, V. R. Pandharipande, K. E. Schmidt, Superfluid Fermi gases with large scattering length, Phys. Rev. Lett. 91 (2003) 050401.

[498] D. M. Ceperley, Path Integral Monte Carlo methods for fermions, in: K. Binder, G. Ciccotti (Eds.), Monte Carlo and Molecular Dynamics of Condensed Matter Systems, Vol. 49, Springer, Berlin, 1996, p. 443.

[499] J. B. Anderson, A random-walk simulation of the Schrödinger equation: H_3^+, J. Chem. Phys. 63 (1975) 1499.

[500] J. R. Engelbrecht, M. Randeria, C. A. R. Sá de Melo, BCS to Bose crossover: Broken-symmetry state, Phys. Rev. B 55 (1997) 15153.

[501] L. Viverit, S. Giorgini, L. P. Pitaevskii, S. Stringari, Momentum distribution of a trapped Fermi gas with large scattering length Cond-mat/0307538.

[502] Y. Ohashi, A. Griffin, Superfluidity and collective modes in a uniform gas of Fermi atoms with a Feshbach resonance, Phys. Rev. A 67 (2003) 063612.

[503] G. Modugno, F. Ferlaino, R. Heidemann, G. Roati, M. Inguscio, Production of a Fermi gas of atoms in an optical lattice, Phys. Rev. A 68 (2003) 011601.

[504] P. de Gennes, Superconductivity of Metals and Alloys, Addison-Wesley, New York, 1966.

[505] J. Ranninger, J.-M. Robin, M. Eschrig, Superfluid precursor effects in a model of hybridized bosons and fermions, Phys. Rev. Lett. 74 (1995) 4027.

[506] J. Stajic, J. N. Milstein, Q. Chen, M. L. Chiofalo, M. J. Holland, K. Levin, The nature of superfluidity in ultracold Fermi gases near Feshbach resonances Cond-mat/0309329.

[507] M. L. Chiofalo, M. P. Tosi, N. H. March, Model of r-space boson-fermion mixture and its relevance to high-T_c cuprates, Phys. Chem. Liq. 37 (1999) 547.

[508] B. Bucher, P. Steiner, J. Karpinski, E. Kaldis, P. Wachter, Influence of the spin gap on the normal state transport in $YBa_2Cu_4O_8$, Phys. Rev. Lett. 70 (1993) 2012.

[509] P. Devillard, J. Ranninger, Pseudogap phase in high-T_c superconductors, Phys. Rev. Lett. 84 (2000) 5200.

[510] H.-C. Ren, An analytical approach to the pseudogap in boson-fermion model and its possible relevance to cuprate superconductors, Physica C 303 (1998) 115.

[511] J. C. Campuzano, H. Ding, M. R. Norman, H. M. Fretwell, M. Randeria, J. M. A. Kaminski, T. Takeuchi, T. Sato, T. Yokoya, T. Takahashi, T. Mochiku, K. Kadowaki, P. Guptasarma, D. G. Hinks, Z. Konstantinovic, Z. Z. Li, H. Raffy, Electronic spectra and their relation to the (pi, pi) collective mode in high-T_c superconductors, Phys. Rev. Lett. 83 (1999) 3709.

[512] V. M. Loktev, R. M. Quick, S. G. Sharapov, Phase fluctuations and pseudogap phenomena, Phys. Rep. 349 (2001) 1.

[513] J. M. Singer, M. H. Pedersen, T. Scheider, H. Bech, H.-G. Matuttis, From BCS-like superconductuvity to condensation of local pairs: A numerical study of the attractive Hubbard model, Phys. Rev. B 54 (1996) 1286.

[514] S. R. White, Density matrix formulation for quantum renormalization groups, Phys. Rev. Lett. 69 (1992) 2863.

[515] S. R. White, Numerical renormalization group for finite Hubbard lattices, Phys. Rev. B 45 (1992) 5752.

[516] G. M. Kavoulakis, G. Baym, Rapidly rotating Bose-Einstein condensates in anharmonic potentials, New J. Phys. 5 (2003) 51.1.

[517] T.-L. Ho, Bose-Einstein condensates with large number of vortices, Phys. Rev. Lett. 87 (2001) 060403.

[518] G. Baym, C. J. Pethick, Vortex core structure and global properties of rapidly rotating Bose-Einstein condensates Cond-mat/0308325.

[519] D. H. J. O'Dell, S. Giovanazzi, G. Kurizki, Rotons in gaseous Bose-Einstein condensates irradiated by a laser, Phys. Rev. Lett. 90 (2003) 110402.

[520] L. Santos, G. V. Shlyapnikov, M. Lewenstein, Roton-maxon spectrum and stability of trapped dipolar Bose-Einstein condensates, Phys. Rev. Lett. 90 (2003) 250403.

[521] L. V. Butov, Exciton condensation in coupled quantum wells, Solid State Commun. 127 (2003) 89.

[522] P. L. Knight, E. A. Hinds, M. B. Plenio (Eds.), Practical Realization of Quantum Information Processing, Phil. Trans. R. Soc. 361 (2003) 1321.

[523] W. F. Vinen, Quantum turbulence, Physica B 329 (2003) 191.

[524] E. Kozik, B. V. Svistunov, Kelvin wave cascade and decay of superfluid turbulence, Phys. Rev. Lett. 92 (2004) 035301.

[525] M. Tsubota, T. Araki, S. N. Nemirovskii, Dynamics of vortex tangle without mutual friction in superfluid ^{4}He, Phys. Rev. B 62 (2000) 11751.

[526] G. E. Volovik, Superfluid analogies of cosmological phenomena, Phys. Rep. 351 (2001) 195.

[527] L. J. Garay, J. R. Anglin, J. I. Cirac, P. Zoller, Sonic analog of gravitational black holes in Bose-Einstein condensates, Phys. Rev. Lett. 85 (2000) 4643.

[528] C. L. Lopreore, R. E. Wyatt, Quantum wave packet dynamics with trajectories, Phys. Rev. Lett. 82 (1999) 5190.

[529] G. de Fabritiis, S. Succi, P. V. Coveney, Electronic structure calculations using self-adaptive Voronoi basis functions, J. Stat. Phys. 107 (2002) 143.

Elenco dei volumi della collana
"Appunti"
pubblicati dall'Anno Accademico 1994/95

GIUSEPPE BERTIN (a cura di), *Seminario di Astrofisica*, 1995.

LUIGI AMBROSIO, *Corso introduttivo alla Teoria Geometrica della Misura ed alle Superfici Minime*, 1997 (ristampa).

CARLO PETRONIO, *A Theorem of Eliashberg and Thurston on Foliations and Contact Structures*, 1997.

MARIO TOSI, *Introduction to Statistical Mechanics and Thermodynamics*, 1997.

PAOLO ALUFFI (a cura di), *Quantum cohomology at the Mittag-Leffler Institute*, 1997.

GILBERTO BINI, CORRADO DE CONCINI, MARZIA POLITO, CLAUDIO PROCESI, *On the Work of Givental Relative to Mirror Symmetry*, 1998

GIUSEPPE DA PRATO, *Introduction to differential stochastic equations*, 1998 (seconda edizione)

HERBERT CLEMENS, *Introduction to Hodge Theory*, 1998

HUYÊN PHAM, *Imperfections de Marchés et Méthodes d'Evaluation et Couverture d'Options*, 1998

MARCO MANETTI, *Corso introduttivo alla Geometria Algebrica*, 1998

AA.VV., *Seminari di Geometria Algebrica 1998-1999*, 1999

ALESSANDRA LUNARDI, *Interpolation Theory*, 1999

RENATA SCOGNAMILLO, *Rappresentazioni dei gruppi finiti e loro caratteri*, 1999

SERGIO RODRIGUEZ, *Symmetry in Physics*, 1999

F. STROCCHI, *Symmetry Breaking in Classical Systems and Nonlinear Functional Analysis*, 1999

ANDREA C.G. MENNUCCI, SANJOY K. MITTER, *Probabilità ed informazione*, 2000

LUIGI AMBROSIO, PAOLO TILLI, *Selected Topics on "Analysis in Metric Spaces"*, 2000

SERGEI V. BULANOV, *Lectures on Nonlinear Physics*, 2000

LUCA CIOTTI, *Lectures Notes on Stellar Dynamics*, 2000

SERGIO RODRIGUEZ, *The scattering of light by matter*, 2001

GIUSEPPE DA PRATO, *An Introduction to Infinite Dimensional Analysis*, 2001

SAURO SUCCI, *An Introduction to Computational Physics. Part I: Grid Methods*, 2002

DORIN BUCUR, GIUSEPPE BUTTAZZO, *Variational Methods in Some Shape Optimization Problems*, 2002

EDOARDO VESENTINI, *Introduction to continuous semigroups*, 2002.

ANNA MINGUZZI, MARIO TOSI, *Introduction to the Theory of Many-Body Systems*, 2002.

SAURO SUCCI, *An Introduction to Computational Physics. Part II: Particle Methods*, 2003

ANNA MINGUZZI, SAURO SUCCI, FEDERICO TOSCHI, MARIO TOSI, PATRIZIA VIGNOLO, *Numerical methods for atomic quantum gases*, 2004

Fotocomposizione "CompoMat" Loc. Braccone, 02040 Configni (RI), Italy
Finito di stampare per conto della "CompoMat" dalla Nuova Grafica 86 nel luglio 2004

Anna Minguzzi
Scuola Normale Superiore
Piazza dei Cavalieri, 7
56100 PISA, Italy
minguzzi@sns.it

Sauro Succi
Istituto per le Applicazioni del Calcolo, CNR
Viale del Policlinico, 137
00161 ROMA, Italy
succi@iac.rm.cnr.it

Federico Toschi
Istituto per le Applicazioni del Calcolo, CNR
Viale del Policlinico, 137
00161 ROMA, Italy
toschi@iac.rm.cnr.it

Mario Tosi
Scuola Normale Superiore
Piazza dei Cavalieri, 7
56100 PISA, Italy
tosim@sns.it

Patrizia Vignolo
Scuola Normale Superiore
Piazza dei Cavalieri, 7
56100 PISA, Italy
vignolo@sns.it

Numerical methods for atomic quantum gases

Key words: Bose-Einstein condensation, degenerate Fermi gases
PACS: 03.75.Fi, 05.30,Jp